TWELVE CITIES

TWELVE CITIES

JOHN GUNTHER

HARPER & ROW, PUBLISHERS

NEW YORK, EVANSTON, AND LONDON

1817

CONTENTS

8   CONTENTS

FOREWORD

This book resembles to a degree its fellows in the Inside series —*Inside Europe,* which was first published as long ago as 1936, *Inside Africa, Inside Russia Today, Inside South America,* et cetera. There have been eight Insides so far, covering the known political world of today segment by segment on a continental basis. The present book differs from the others mainly in that it deals with a group of major cities, not continents en bloc or a single great country like the United States.

What I have tried to do above all is to describe with some diligence, scope, and detail what these twelve cities scattered over the world, each representing a different country, are *like* —to give a picture, an estimate, of their mood, color, texture, tempo, distinctions, problems, perturbations, implausibilities, as well as to deal with the structure of their politics and government.

Such questions as these arise:

What makes London great?

Why did Paris explode in the spring of 1968?

What's wrong with Rome?

How are cities in the Communist bloc governed on the municipal level?

Why is Brussels the victim of an acrid language crisis?

What are some of the constituents of Viennese *Gemütlichkeit?*

What makes power in Beirut?

Who runs Tokyo?

Today is the age par excellence of great cities—huge and voracious cities proliferating all over the place, seeping like ink-blots into the countryside, flashing like prisms, and linking themselves to other communities to form the new type of boundless urban metropolis. The growth, density, and seminal energy of these human mulligatawnies is, of course, one of the overriding phenomena of this exacerbated century—stimulating too. Almost every type of issue enters into the problem of the contemporary megalopolis—urban planning, the curse of the automobile, education, housing, public entertainment, segregation, the slums, unemployment, public health, smog, water supply—to say nothing of savage riots on hot dirty streets. We in the United States already have anguished knowledge of most of this, even if we don't seem to be doing much—or enough—about it. So it may be useful to turn to some of the great capitals of Europe and Asia as a matter of comparison—perhaps for enlightenment as well.

I took four different trips for research on this book—to the Far East, London, Western Europe, and Eastern Europe plus the Middle East—and it has been several years in the making. The spark that set me off was a remark by Hobart Lewis, president and executive editor of *Reader's Digest,* who suggested casually in Chicago in 1966 that, having done profiles of continents, countries, and personalities, I might try my hand at something different—cities. The *Digest* made my expeditions possible, let my wife come along, and printed several articles which were condensations of parts of the present text. My thanks are profuse. *Harper's Magazine* also gave hospitality to two articles on London. But all this first-run material has now been substantially revised, reworked, rechecked, and augmented to make the present book. The *Digest* articles seldom ran to more than half a dozen pages each, but here several chapters are much longer—running to ten thousand words or more.

So—we begin with London, the king city of them all.

TWELVE CITIES

1. THE NEW OLD LADY
ON THE THAMES

I fell in love with London at first sight when I arrived there about 4 P.M. on a gray silken day in the late spring of 1922, and have been in love with it ever since. I was a twenty-year-old student at the University of Chicago at the time, taking my first trip to Europe on a cattle boat. During the long procession of later years I have lived in London for several periods, visited it on numerous occasions, and seen it in a good many different moods and stances—during the rancid agony of unemployment in the thirties, the prewar interval of national dilemma and hesitation, the heroic war years themselves, and alternating times of well-being and depression ever since.

The first series of articles I ever wrote for the Chicago *Daily News*, when I was its assistant—and extremely junior—London correspondent in 1924, was on the subject, "Is London Finished?" This was ordered by our Chicago editors as a result of some pessimistic remarks on England's future by an American ambassador of the time. People ask the same kind of silly question today. But London is still there.

Indeed it is. This stupendous capital is, to my mind, the greatest metropolis in the world. It isn't as old as Rome, as luminous as Paris, as spectacular as New York, or as big as Tokyo. It has all manner of negative qualities, like the weather (ever seen a true West End Londoner without an umbrella?), its provincialisms (like the licensing hours in pubs), and its archaic preoccupation with class, even in the "lower" classes; nobody can be more of a snob than a true Cockney. But I will report changes in all these fields before this chapter and the next are done.

Nevertheless, again in my own view, London is the greatest of all cities for several cogent reasons. It has grace, durability, style, good manners, and, above all, formidable weight. It knows what true satisfactions are—even though the thermometer in the men's room of a good club stood at 53° Fahrenheit the last time I was there. Its richness and diversity are multiform. At once it is intricate, ponderous, spirited, and kind.

The words of Dr. Samuel Johnson are well known—"When a man is tired of London, he is tired of life; for there is in London all that life can afford." And a line from Alexander Pope is equally appropriate—"Dear, damn'd, distracting town."

Another quotation I like comes from Marshal Blücher, the Prussian who helped beat Napoleon at Waterloo. Gazing at London for the first time from St. Paul's after the great victory Blücher muttered gruffly, *"Was für Plunder!"* ("What loot!")

London has changed a good deal in the past few years, so much so that a "new" metropolis seems to be arising. Even the structure of government has become radically different since April 1, 1965. The London County Council, for several generations probably the most celebrated instrument of municipal authority in the world, has given way to a new administration called the Greater London Council, which we shall inspect a few pages farther on.

Changes have taken place in almost every dimension—architecture, street scenes, the attitude to Queen and country, food, the look of the people (it's better, particularly their teeth), the social structure, urban development, clothes, and youthful folkways. This huge conurbation (a fashionable new British word meaning "the coming together of built-up areas") is not standpat, but distinguished by evolution, experimentation, change.

But first let us glance at this magnificent and copious city in the large, before describing its metamorphoses in detail or discussing its unique form of government, which is changing too.

Backdrop to the Conurbation

London, a triple capital (of England, the United Kingdom, and, in a manner of speaking, the British Commonwealth), has a population slightly over eight million, which puts it second among the world's cities, a bit ahead of New York, but behind Tokyo. The area of Greater London is 620 square miles, almost twice that of New York City (319.8 square miles). The Thames bisects it in a series of loops like an intestine. It contains roughly a third of all the wealth in the British Isles, and holds one-sixth of the total population of Great Britain. One startling statistic I came across is that about 10 percent of all London births are illegitimate.[1]

I don't know any city which gives such an impact on being approached from the air. The spectacle is solid, immense, and seemingly continuous, particularly at night. Orange beacons, like illuminated pumpkins, mark the long roads and avenues, and smaller lights flash on thousands of lesser streets, scarcely one of which is straight. Even the shortest lane seems to have a curve, bend, bow, arc, hook, or jog. This is a city built in the pattern of ellipses, horseshoes, crescents, and parabolas; no rectilinear gridiron encases it.

Nobody knows for sure what the word "London" means. Its earliest inhabitants were Celts. "Llyn-din," a word in the old Celtic language, means "a fort adjacent to water"; this may have been the origin of Londinium, the Roman name, but most contemporary authorities think not, without being able to offer a better derivation. The city rose where it did, on the banks of the Thames forty miles from its estuary, because this was a convenient strategic location. Historians tell us that the site was dominated by two small hills, and here was the first point where the river, which has been nicely described as "liquid history," was narrow enough to be forded or bridged. And the sea, with its pregnant opportunities for trade, was close.

[1] The over-all rate for the United States is about 8 percent.

The Roman conquest began with Julius Caesar in the first century B.C., as almost everybody knows, and lasted about five hundred years. One early national heroine, Queen Boadicea, fought to beat the Romans back; but she was beaten in battle, and killed herself. London likes its folklore, even the hoariest, and one persistent legend is that Boadicea's body still lies today beneath platform No. 10 in King's Cross Station.[2]

The Romans built roads, like Watling Street, now Cannon Street, leading to what are now the channel ports; the paths of these are still followed mile after mile by highways today. Elephants trod these roads in ancient days. Minor Roman remains are still found regularly, and as recently as 1965 important artifacts, dating back more than two thousand years, were unearthed in excavations near St. Paul's. After the Romans came Angles, Jutes, Saxons. Middlesex, the historic London county which has recently been administratively abolished and written off the maps, means "Middle Saxon." And then came Danes, Vikings, Normans. William the Conqueror was crowned in Westminster Abbey in 1066, and London, with a population of about fifty thousand, became the capital of his realm. A previous capital was Winchester.[3]

The history of London as the heart of England has never been challenged subsequently. Westminster (western ministry) has been the site of Parliament since the thirteenth century, and the Archbishop of Canterbury has had his headquarters in Lambeth Palace across the river for more than six hundred years. The population rose to something like 180,000 in Queen Elizabeth's time, and she took stern measures—reminiscent of several being tried today—to control the expansion of the metropolis. The Great Plague came in 1665, the Great Fire in 1666. London had to build from a new beginning; even so, the population jumped to 850,000 by 1800. Pall Mall was the first street in the world to be lit by gas (1805), one authority relates. Another curiosity is that railways got under way only slowly because, so the story

---

[2] *London, a Souvenir in Pictures,* p. 12.
[3] Christopher Trent, *Greater London,* pp. 23-25.

goes, the Duke of Wellington thought that they would facilitate invasion of the country by the French. Now, of course, London is the essential pivot of a complex and comprehensive national railway system, and France and England are to be connected soon by a channel tunnel.

Today this colossus of cities is the third port in the world (after New York and Rotterdam), as well as one of its greatest financial, industrial, mercantile, and trading centers. But it is rich in a good many other directions as well. Few other cities have given so many geographical metaphors to the language—Downing Street, Rotten Row, Fleet Street (which was once a river), Whitehall, Harley Street, Billingsgate, Petticoat Lane, Savile Row, and London Bridge, which, following the prophecy of the old rhyme, *is* actually falling down today—more precisely, slipping farther into the muck of the Thames a few inches every year —and which was bodily sold to interests in the United States in 1968. Names associated with London have become fixed in common speech all over the world—like Sherlock Holmes, Madame Tussaud's, Old Vic, Mother of Parliaments, Soho, Wimbledon, Bloomsbury, Mr. Pickwick, Covent (originally "Convent") Garden, Big Ben, Shepherd Market, Greenwich Time, and Scotland Yard.

Street names are a revealing study, as anybody may find out by reading a small book, *Street Names of the City of London,* by Eilert Ekwall. In the financial district alone one may find Knightrider Street, part of which was once called Old Fish Street, Royal Mint Street (formerly Hog Street), Stinking Lane (which became King Edward Street in 1843), Seething Lane, Hercules' Pillar Alley, Bearbinder Lane, and Stockfishmonger Row. In other parts of London are Greatorex Street, All Hallows, Iron Gate Wharf Road, and Newington Butts. There are a number of Alexandras (avenue, gate, and two roads), Cambridges (circus, grove, park, path, gardens, terrace, and two roads), and Chelseas (barracks, basin, bridge, bridge road, creek, road, manor gardens, park gardens, reach, and square). Pity the taxi driver!

Then, too, London folklore is voluminous, as well as sedulously kept alive. Every night an armed detail, known as a picquet, of the Guards marches to the Bank of England, the Old Lady of Threadneedle Street, to stand duty there, although such protection is altogether unnecessary these days, and is still preceded by a lantern carrier, because this is the way it was in days gone by. Yeomen of the Guard still search ceremoniously for a modern equivalent of Guy Fawkes in the cellars of the House of Lords every time Parliament is convened, a tradition which dates from the Gunpowder Plot in 1605. Six pinioned ravens still prowl near the Tower of London, because a medieval legend proclaimed that the British imperium would end if the ravens which clung to the Tower (like the pigeons in Trafalgar Square) ever went away. As a matter of fact, the tradition of the ravens has, it turns out, outlived the Empire. A guidebook tells us that the ravens are kept by a "Yeoman Warder," and each receives a weekly allowance of 2s 4d worth of horseflesh a week.

Precisely at 9:53 every night—so I have learned—the principal warder of the Tower advances with a candle holding two golden keys aloft. These are turned over to the guard. The sentry commands, "Halt! Who goes there?" The warder replies, "Queen Elizabeth's keys, all is well." This ceremony has not been omitted any night for more than four hundred years, even during the hazards of the Blitz.[4]

✿

For several generations in the modern period London was a socialist metropolis, run by a Labour majority in the municipal administration, although several of the individual boroughs have usually had Conservative administrations and the impregnable inner square mile known as "the City," the financial district (of which more later), is an entity strictly on its own. An election in April, 1967, brought a Conservative majority to the Greater London Council, a formidable upset. But, even while its government was Socialist, London stood out as one of the most indestructible

[4] *Christian Science Monitor*, August 17, 1965.

surviving symbols in the world of the power of private commercial interests, private wealth, and private property.

Who owns London? This is a vastly complicated subject because of the leasehold system; ownership is inextricably mixed up with long-term tenantry. By the census of 1961, 36 percent of all tenancies were held by "owner-occupiers," to the number of 964,-943, but I was never able to get anybody to explain to me satisfactorily what an "owner-occupier" is, since he doesn't own. Large agglutinations of London property are still, however, concentrated in a few outright individual holdings. The Duke of Westminster owns 300 acres in the Piccadilly–Bond Street area and near Grosvenor Square. The Portman Family Estates have 100 acres north of Oxford Street, and Lord Howard de Walden is the owner of 100 acres near Regent's Park. The Cadogan estates cover 90 acres in Chelsea. The Duchy of Cornwall owns parts of Kensington, and the Bishopric of Ely owns the area near Hatton Gardens, the celebrated diamond market. The City, i.e., the financial district, has large properties on Bond Street and in adjacent neighborhoods. Several Oxford colleges, notably All Souls, which has holdings in the Jermyn Street area, derive handsome revenues from their London properties—in fact, live on them.

In theory the Labour government of the day—Mr. Wilson's government at the moment—deplores inequality in the distribution of land in London, but there is little it can do to change it. Big landlordism has, however, its advantages, which even Labour ministers concede. First, the large owners tend to keep up their property better than the small. Second, extensive holdings make it easier to plan new developments. Another point is that the municipality itself has extremely substantial properties. The Greater London Council is, in fact, the largest landlord in the world.

In today's insufferably crowded and complex world London offers virtually every service. You can buy anything from a secondhand bus to witch balls or a treen, even including a knife with 1,851 blades.[5] Every child is entitled to free milk every day,

---

[5] I cannot track down the source of this detail, but I believe it to be true.

and about 80 percent of the children take advantage of this. The telephone operator will wake you in the morning if your alarm clock isn't working, and, if you are out of change (but possessed of a post-office telephone credit card), you can telephone on credit from any call box. An exchange called "ASK" is prepared to answer various questions, and by dialing "CHICKEN" you can have a hot meal sent over. Bookshops are the best in the world. Mail is delivered in metropolitan London twice a day, and the efficiency of the postal service puts ours to shame. The post office will provide you with a messenger for almost any service. Ash cans are padded to make them less noisy. A Gamblers Anonymous exists on the pattern of Alcoholics Anonymous, and there is a Suicides Anonymous as well. A bank in the City stays open Saturdays and Sundays for cashing checks.[6]

London fosters the amenities: people take pleasure in the art of living, and men and women of station are expected to be civilized. Theaters open early, so that there is ample time for supper afterward. You can get into a taxi without breaking your neck, and the back windows of most have a bluish glaze, which blocks the view of Peeping Toms. (But, for all its addiction to discretion, London has more publicly known and noisy amatory scandals than any comparable city in the world.) Whiskey and gin are only 72 proof, which means that Londoners have to drink more than New Yorkers to get drunk. (But I have seen plenty of wild London drunks.) London has incredibly vexing confusions as well. Kensington lies on the north side of the river, but is in the district called Southwest One, Southwest Seven, and West Eight. And there are plenty of nonamenities such as that the buses and tubes don't run all night, and people are obliged to stand eternally in queues even for staple groceries. Housewives are unable to buy large quantities of perishable foods at a time because their refrigeration facilities are apt to be inadequate. Central heat, where it exists at all, is painfully feeble, and most

[6] *The New London Spy*, edited by Hunter Davies, a wonderful compendium.

Londoners have chilblains and suffer actively from the wettish cold every winter.

London has a splendid patina, respect for law, resolution (recall the Blitz), and good talk at dinner parties, even the stiffest. Dinner parties are a story in themselves. Guests are astonishingly mixed; bitter political opponents may sit side by side. An accomplished hostess will do her best to collar an antediluvian bishop and an avant-garde sculptor for the same table. But guests are seldom identified or even introduced; nobody can be more neglected than a stranger at a London cocktail party. One thing I like is the average citizen's intense regard for punctuality, one of the true bases of good manners. Dinner invitations may still read "Seven-forty-five for eight," meaning that guests will be given a quarter-of-an-hour leeway in arriving, but not more. One peculiar device is the small wooden easel, with movable blocks naming members of the family, which is placed in the entrance hall to indicate to a caller who is home. Another well-kept tradition is that the footman, if you still have one, whispers to each guest the name and vintage of the wines being served.

Engagements are made far in advance—unbelievably so to an American. I know one ambassador of a foreign country resident in England who, in February, already had all his summer weekends booked. Recently a London publisher wrote to ask an American friend who was about to visit London for dinner on March 17; the American, after accepting, was obliged to inform the Londoner that his plans had been changed and that he could not be in London until October. Instantly (in March) he received a cable: "Dine with us October 3 *unless already engaged.*"

Clubs are a subject so much written about that I am inclined to skip them. Everybody knows about the learned quality of the Athenaeum, the camaraderie of Buck's, the esoteric distinction of the Beefsteak, and the glacial toniness of White's (for which the waiting list is eight years long), Brooks's and Boodle's. The gist of club life is, of course, that it gives men a haven, a refuge, from their wives, as well as an address away from home.

London Changing

First, the skyline. Until a few years ago, except for its constellation of majestic public buildings in a limited area, London stood from two to four stories high. This is still true of the metropolis in bulk, but there has come a sharp change in the central regions. The new post office building, with its television tower, rises 579 feet near Tottenham Court Road, and you can have dinner in a restaurant on top. St. Paul's, no less, has become encased on three sides by large glass office buildings, and a new concrete skyscraper in the most modern style rises next to Westminster Abbey, dwarfing it. Milbank Tower, a new office building on the Thames, is thirty-two stories high, and dumpy old Victoria Street has been reborn into silverish glassy blocks.

Old London had hundreds of thousands of chimney pots, one for each fireplace or flue in a house. New London has few. Old London had extensive dreary slums, particularly in the East End, but these are being systematically—if slowly—renovated. In part this has been made possible because several of the former major slum areas were blitzed out.

Another change obvious at a glance is what might be called continentalization. The town is full of French, Italians, Spaniards, and you overhear strange accents on the streets—whiffs of exotic languages. One veteran American resident of London told me that this is the single change that strikes him most. Many Central European Jews, with their intensely civilizing influence, settled in London after the war. Soon came a wave of West Indians, and mixed neighborhoods rose in districts like Notting Hill; London had its first race riot in 1958, and racial tension is an angry and rising problem. Later a new invading swarm arrived—the Cypriotes, who, among other things, have taken over innumerable restaurants usually called Greek. Somewhat intemperate, they have been the source of a good deal of minor crime and violence.

Food has changed. London used to be a grim, dreary place to

dine in unless a visitor knew exactly where to go; the range of restaurants was limited, and those really good were fabulously expensive. But now gay small restaurants with checkered table-cloths in the French manner have sprung up almost everywhere, and serve exciting continental fare. Some, hidden in alleys, are hard to find, and, from the outside, look like dungeons; but they gleam within. Waiters are not of the waiter "class"; they wear blue jeans and long hair, and look like sons of the proprietor—maybe they are. Then, too, steak houses, coffee shops, and "Wimpy Bars" have cropped up everywhere. One chain of restaurants is called "Old Kentucky," and one place bears the name "Dunkin' Donut"—minor instances of the Americanization of London which, in certain spheres, goes on steadily.

Night life has expanded, too. London is not a wide-open town and never will be, because of the national character, and the prostitutes have been chased off the streets. But you can go to "clubs" where girls in strip-tease shows peel down almost to the altogether; I visited one—in dear old London!—where a girl un-dressed in the company of a live cheetah. Completely naked girls, however, are not allowed to move their bodies in a per-formance; they stand stiff as wax effigies. Many joints and dives in Soho are called "clubs" because this dodge appeals to British snobbery, serves to keep the police out, and makes it possible to sell drinks after hours. Most of the chic discotheques are clubs. A few, on the level, really *are* clubs, with restricted membership, but you can tip your way into most, and some are just about as exclusive as the telephone book. The discotheques range from the grandly immaculate to the coarsest grubby; one or two, like the Sibella, are as brilliantly sophisticated as any night clubs in the world.

In July, 1960, the new Betting and Gaming Act received the royal assent, and what are known as "betting" shops became legal. Londoners notoriously love the horses, and there are more than sixteen thousand such shops in England now. More than a *billion* pounds (almost $3 billion—before devaluation) were spent last year in gambling on football, greyhounds, and the track. It was interesting, ducking into one of these tawdry betting "shops," to learn that TV and sound broadcasts are forbidden on the premises; results are announced by a loudspeaker. This is to discourage loitering. You bet, lose or win, and get out. I saw a crude sign, "Clients must write out thier [sic] own bets." At the other extreme are the fashionable gambling clubs, like Crockford's and Aspinall's. This latter, on Berkeley Square, is housed in one of the most exquisite old buildings in London, with a William Kent staircase marvelous to behold; and it was moderately interesting to learn that an American millionaire had lost $57,000 in two hours a few nights before at the very table where I was standing. At another place I ran into a game of *chemin de fer* which had gone on uninterruptedly for twenty-eight hours; relays of exhausted croupiers served the players. Havens for the rich like these are fitted out with an almost excessive sumptuousness and luxury, like sets in the movies. At the other end of the line are smaller, less exclusive establishments, some with discotheques attached, where you can play roulette for shilling stakes. But a thimble-sized dry martini costs about $2.

Even the weather has changed. London has plenty of bad weather still, but real pea-soupers, when the city lies dead and blind for days on end under greasy yellow billows of acrid fog, seldom occur these days. This is largely because of the mandatory use of smokeless fuel, not only by industry but in the millions of old fireplaces that lead up to the millions of old chimney pots. Twenty years ago a white collar worn in the streets became black in an hour, and the fog often made it impossible to see more than a block or two. Now visibility is better. Of course it still does rain a lot (but not much more than in New York, as a matter of fact), and sunshine is notably sparse most of the year. The most irritating thing about rain in London is that it doesn't come down in sudden torrential spurts, as it often does in New York, but makes a steady drizzle. The eternal grayness is what makes the city so depressing to many—that and its chill. Maybe the freezing moisture helps to give Londoners their famously rosy cheeks, but most of those I know would welcome a climate less rigorous.

There are any number of changes in minor realms. The best thing to read in London, the personal or "agony" column of the *Times*, appears on page two instead of one, now that the *Times* has sacrificed its first page to news, and there is a temptation to forget to look for it. Decimal currency is on the way, which will necessitate a vastly complicated metamorphosis in vending machines and parking meters as well as in many other ways. Scarcely anybody wears white tie any more except at the most rigidly formal functions. You can buy bottled liquor nowadays outside licensing hours. But a good many traditional characteristics have *not* changed. You can still, as an example, buy a shirt at a good shop that will last five, six, seven years, and not show a sign of wear. Hostesses still proudly display on the mantelpiece the cards of invitation they have received for coming events, even though this is supposed to be a reticent city.

Servants are becoming almost as hard to find as in New York, and taxis are maddeningly scarce in certain districts at certain hours, as they now seem to be everywhere else in the known world. This comes, awkwardly enough, at a period when more and more middle-class Londoners are, for the first time in their lives, able to afford such luxuries as servants and taxicabs. England as a whole is in the grip of a severe economic freeze, but, a strange paradox, London itself gives a good many external signs of being prosperous. Partly this is because unemployment is being held at a minimum.

Coupled with a substantial and perhaps ill-based euphoria is a certain amount of spiritual malaise, particularly among the middle-aged. Indirectly this malaise may be due to the loss of Empire and the departure of the old imperial hegemony. London had a central myth which held it together. Even the humblest citizen had a pervasive pride about the British position in the world. There were other powerful elements making for cohesion —such as the Royal Family, the City, the Church, the civil service, and the Establishment in general. These are still powerful and persuasive forces, but almost each has undergone change in one way or other. The sister of the Queen of England married a photographer. Twenty years ago nobody could get a job with the BBC without an Oxford accent; that swift speech, with the somewhat effeminate lisp, became famous as a kind of parody of English speech all over the world. Now the BBC hires any kind of accent.

All this brings up the difficult subject of class, where, again, changes have been profound. The aristocracy can't afford to live as it once did, which tends to level society off. I know Londoners not particularly wealthy who grew up thirty years ago in houses where a staff of eight or even more servants was taken for granted. Young men in fashionable regiments were—only yesterday—subject to stringent rebuke or even dismissal if, after hours, they were caught wearing anything but a bowler hat and stiff white collar, or were seen on a street smoking a cigarette or carrying a bundle. The whole social structure has been loosened up, or, as a friend of mine put it nicely, "loosened down."

A top layer of society certainly still exists, and London is a city where thousands of citizens still "know their place." But their "place" has become more secure and more of their own choosing. Pubs still have two entrances: one for the squire, one for the working man (and there are no fewer than six thousand pubs in London), but, fascinatingly enough, the working-class division is apt to be more jealous of the inviolability of its sanctuary than the other. The old-school tie certainly isn't gone, and the Foreign Office and City still recruit largely from the public (i.e., private) schools. There are, in effect, two educational systems in the country, one for the privileged and one for the nonprivileged, a deleterious and dangerous situation, but thousands of young men and women are cutting across these and other barriers now. Dominant parents no longer choose professions for their sons in the Victorian manner, and Eton boys take jobs—like hauling mail —on Christmas holidays. The "red-brick" universities are flourishing.

What has contributed more than anything else to smoothing

down the caste barrier is the trend toward equalization of income and opportunity. If taxation takes the rich man's wealth away, it becomes difficult for him to remain superior.

It is still true that if a duke gets drunk and falls out of a window he will be front-page news, which is not the case with Jack Robinson, his servant. But dukes and duchesses are, in a manner of speaking, dated. Talent is what counts these days, not social rank. Nobody gives a damn if you are a "gentleman" or what kind of English you speak if you have talent and become a success. One indication of this is that a whole new set of folk heroes has emerged, like the Beatles and the new generation of actors and movie people. Peeresses caught in adulterous beds still get lively headlines, but the publicity emphasis is all on new plebeian talent, which is spirited, vigorous, and omnipresent.

Finally, more than ever before, London is being constantly fructified these days by outsiders, particularly from the North— Prime Minister Wilson is a Yorkshireman, and most of the new generation of style specialists, TV executives, newspaper owners, industrial tycoons, and particularly movie stars, come from the Midlands and the North. One principal spawning ground for new talent is Liverpool. This city was mercilessly punished by the Blitz, and, dull as it is, its ashes became a kind of fertilizer, a hothouse for off-beat creative energy.

Youth, Youth, Youth

We come now to what became famous as "Swinging London" two or three years ago. This was a misnomer, because only a small element of youthful London ever did much swinging. Most Londoners profess to be bored by the subject, and tend to dismiss it as an aberration. But they cannot deny that the youngsters gave parts of the metropolis a new look, and made London the brightest city in Europe for a period. Berlin and Paris are still old-hat, even staid, by comparison. What a reversal of roles this was! London became the city of fleshpots, and Paris the place where nobody talked about anything much but the price of automobiles.

The sociologists tell us that the London swingers descended from two rival groups, the Mods (Moderns) and Rockers. The Mods went in for fancier dress, looked effete, and rode on motor scooters if they didn't walk; the Rockers, tougher, dressed mostly in leather, wore white helmets with big goggles, and tore around on motorcycles. They thought that the Mod boys—with their long hair, 5½-inch-wide patent-leather belts, and spangled shirts— were sissies. Some Mods even carried handbags and used lipsticks, but, contrary to the usual notion, they were not predominantly homosexual. Their dandyism derived partly at least from a healthy desire to attract the other sex. To be a queer in London today—outside the worlds of interior decoration, the theater, and the ballet—is to be thought square by most of the young, although there are thousands of homosexuals who are accepted by almost any company.

The spectacle presented by a good many young Londoners is still electrifying. They speak a slang of their own—"bird" for girl, "gig" for party, "nosh" for food, "rozzers" for police, "tat" for shop—and swarm into the discotheques, dance with grim frenzy, and make a new place fashionable for a week, then move on to another. They have little interest in public affairs, think that the "angry young men" of the John Osborne epoch are as dated as Moses, and seldom talk politics. They have washed their hands of taking care of the world, or of being the world's bellwether or watchdog. Some who do not live at home have no regular place to sleep and go about with a bedroll, a toothbrush, or perhaps a guitar as well—with which they camp on any friend's doorstep, or in an empty hall.

What mostly distinguishes them is their dress, as a walk down King's Road in Chelsea will amply show. Of course the girls wear miniskirts, or even micro-miniskirts. Whether the miniskirt, which has now become a standard costume all over the world after its birth in London, is an aphrodisiac or not is a moot question, probably dependent on how pretty the girl's legs are. Knees are seldom the most attractive part of the female body. The boys go in for huge collars and drainpipe or pipestem pants, although flared bell-bottom trousers are in vogue as well. In one discotheque I saw a young man who, I thought, must be an actor who had arrived from some Shakespeare performance without changing his costume—a peach-colored velvet jacket with lace ruffs and sequined trousers. But no, this was his "ordinary" wear. Girls in fashionable clubs wear peaked caps, the tightest of tight slacks, and loose wool sweaters. The present custom for boys to wear their hair longer than girls—another London invention—has of course spread all over the world. Feet are still the easiest giveaway if an old-fashioned viewer finds it difficult to tell a boy from a girl.

One commanding figure in the contemporary mode is Mary Quant, a Liverpool girl who launched the miniskirt. Another principal institution is, as everybody knows, Carnaby Street, which lies on the frontier between Soho and Regent Street. Here an enterprising young man named John Stephen set up a group of shops which, capitalizing on the revolt against dowdiness, first catered to the Mods. The Stephen shops, five or six under different names, have huge fashion photographs on the ceiling and kooky mannequins lying on their sides in grotesque positions.

How can the London youngsters afford their evening forays? For one thing most—both boys and girls—earn quite good salaries for Europe, as much as £18 a week, and, since most are young enough still to be living at home when they "live" anywhere, they have little or no rent to pay. Clothes are relatively cheap at the bright new boutiques which are a madly flourishing business all over the place. It is not fashionable to drink much, nor to gamble, and this too saves money. Girls usually pay their share on a date.

At least four factors assisted the rise of the swingers, both male and female: (1) reaction against the conservatism and conventionality of their parents; (2) an impulse to spontaneity and release after long years of austerity; (3) more earning power; and (4) a sense that the world was doomed and that they might as well have their fling while they could. The whole movement is rooted in protest.

Frightened elders, particularly emigrants from the continent, are much shocked by this ball-before-Waterloo atmosphere, and even go so far as to say that London today reminds them of Berlin before Hitler, with its insane and evil vortex of corruption. But contemporary hijinks in London are mild—almost innocent —by comparison. Berlin was overtly vicious, which London is not, and the swingers in their plumage are a sign of vitality, in their own peculiar British way, rather than of decay.

Miscellany

I went to a James Bond film and, leaving the theater, happened to count the trees in Leicester Square, on which the theater abuts. There are only fourteen, in case anybody wants to know. In a few days I encountered several taxi drivers distinctly rude, a new phenomenon for London. I was impressed as always by the terminology of public signs, such as one near Grosvenor Square which says, "A Person In Charge Of A Dog Who Fouls This Footway Is Liable To A Fine of £5." The Indian Pale Ale at the Connaught is a heartening experience. The London theater, TV, radio, and newspapers are the best in Europe, and no shoe has ever been shined until it has been shined at Claridge's.

2. MATTERS OF
GOVERNMENT AND
SUCH IN LONDON

London never had a metropolitan authority at all until 1855, when a makeshift body was set up. Conditions were outrageous. Hundreds of streets had no names; thousands of houses had no numbers. There was no main drainage system, no Thames Embankment to prevent floods. Sewage ran openly into the Serpentine in Hyde Park; noxious gases came out, and London had a serious cholera epidemic as recently as 1860.

In 1888-89 came the formation of the London County Council (LCC) which administered the metropolis (except for the City) until 1965. Under the LCC, London became a new entity, a county of its own, the youngest in England, assembled from Middlesex and other counties. Roughly an analogy would be the transformation of New York City into a new "state" comprising portions of New York State, New Jersey, and Connecticut.

Thus a new phase in London history began with the LCC, which administered 117 square miles—what is known as "Inner London" today—in an area bound by a five-mile radius from Charing Cross. This was divided into an administrative structure of twenty-eight boroughs, each with its own mayor, council, and local administration, under over-all LCC authority. The LCC was undisputed boss.

Its early mood was radical and reformist, strongly influenced by the Webbs, Bernard Shaw, and the Fabians. Although the Labour Party itself scarcely existed in the early days, Labour influences grew more powerful. This was the era of what was called "Gas and Water Socialism," with do-good zealots committed politically to better municipal services for the people. As time went on, the LCC became more representative of the right rather than the left wing of the Labour Party. It was clean but rigid. Party discipline was emphatic, and it became heavily bureaucratized, somewhat dry, and fantastically powerful within its own domain.

The inheritor, the second generation leader, was Herbert Morrison, who ruled London virtually singlehanded from around 1933 to 1940. A remarkable Cockney, he advanced later to hold various high offices of state, including the Foreign Office, and eventually was elevated to the House of Lords as Lord Morrison of Lambeth. Except for a series of accidents he would have been Prime Minister after the war instead of Clement Attlee. His central passion was always London. "Our 'Erbert," as his beloved Londoners called him, was a beguiling character, the son of a policeman; he had only one eye, the result of an accident. His education was rudimentary, and in his youth he worked as an errand boy, a salesman in a shop, and a telephone operator, until he rose to become secretary of the Labour Party. Once, in a fight with the national government over rebuilding Waterloo Bridge, he set out with spade and shovel to dig new foundations for it with his own hands—or at least made a symbolic gesture to this end—causing enough public commotion to force the government to yield.

After World War II it became clear that London was growing so fast that it would have to be reorganized. The LCC was doing a good administrative job, but it was somewhat sterile and its writ did not extend far enough. So, following prolonged study, a new body, the Greater London Council (GLC) was created to replace it, and the historic old LCC disappeared in 1964-65. The frontiers of the metropolis were pushed out to an average radius of roughly twenty miles from Charing Cross, instead of five, so that "Greater London" grew to cover 620 square miles as against 117 under the LCC, with a population of more than eight million as against three and a quarter million. All of Middlesex was swallowed up, together with parts of Essex, Hertfordshire, Kent, and Surrey as well as the three great "county boroughs" of Croydon, East Ham, and West Ham. New frontiers were drawn, with the whole immense megalopolis divided into 32 new boroughs instead of 28; since the area was greater, the new boroughs were much larger than the old, but the number of local authorities was cut from 90 to 34 to increase efficiency.

Politics—good old-fashioned party politics—played a considerable role in all this. The transfer from LCC to GLC was worked out during the dying years of the last Conservative government, and Labour spokesmen denounced the process as a Tory plot to break Labour's seemingly unshakable hold on London. By making London bigger, the Labour forces charged, the Tories were attempting to extend their own power in the metropolis. Many suburban areas, like Surrey, were traditional Conservative strongholds, and adding their votes to metropolitan elections for the new GLC would inevitably dilute the over-all Labour vote, or so it was assumed. The more strength the Tories gained in the suburban areas, which tend to be more conservative (as in the United States) than the heart of the city, the better their chances would be to control London as a whole. The reorganization scheme was, in a word, presented as a Tory plot to grab off the inner citadel. Morrison called out in the House of Lords, "They're murdering our child!"—or words to that effect. But, contrary to several predictions, Labour won the first elections for the new GLC in 1964, with the consequence that the over-all administration of London remained socialist for the time being. But, as has been noted above, new elections in 1967 went conservative by a heavy margin.

One remarkable constituent of the London area is the Green Belt, a "girdle" completely encircling the metropolis. This, a Morrison concept, came into being in the late 1930's, and has been zealously safeguarded ever since. The idea was to give London a lung, to keep the metropolis within bounds, and to prevent urbanization of the countryside. So, uniquely among the world's great capitals, London has a surrounding protective glacis of green open country, 20 miles wide on the average—not a mere ribbon, but an arc covering 842 square miles in all. Operated jointly by the GLC and the local authorities, it has more than 10,000 acres of farm lands and small country estates; hundreds of tenants have properties ranging from an acre up, all within a metaphorical stone's throw of the heart of the metropolis. But now the steady outward push of London brings new problems, and whole new cities are being built *beyond* the Belt. London has been forced to hop over its protective green, as we shall see. But the Green Belt itself remains untouched, and is a pleasant and healthy amenity. Think how much pleasanter the outskirts of Chicago or Buenos Aires would be if they had adopted the same device.

How the Greater London Council Works

As extraordinary as anything about London is the fact that scarcely anybody knows the name of its titular head of government. I searched around for days before finding out. In a manner of speaking he is the equivalent of John Lindsay in New York or Richard J. Daley in Chicago, but not one Londoner in a hundred thousand has ever heard of him. Officially his title is not "Mayor" but Chairman of the Greater London Council, and his term of office is restricted to a year. Largely his functions are ceremonial —with no politics permitted—a startling difference from the practice of most other great cities of the world, particularly in America.

The incumbent when I was doing research for this chapter was the Rt. Hon. Herbert Ferguson, aged sixty-two, a chartered accountant born in Croydon. Since 1927 he has been associated with a civil-engineering and contracting company—no fancy heritage here, no "class." Having joined the Labour Party in 1945, he devoted many years to municipal affairs. Tall, lean, with silver hair and the ruddy cheeks that are a London trademark, with a people's accent, Mr. Ferguson is alert and worthy. He took my wife and me to lunch at Royal Festival Hall, the superb concert house which is one of the principal monuments of the municipality. With us came Sir William Hart, another tall lean man in his early sixties, with steel-colored hair, an unlined face, and a literate and sophisticated background; he was an Oxford don before becoming clerk to the GLC. He is the permanent administrative arm of the London government under the titular leadership of the chairman. Education, traffic, and housing are their principal preoccupations. As an example, Greater London will need 550,000 new housing units by 1980. The conflict—not between these two men, who work as a team, but at large—is over what to favor most, the man or the machine. More parks or more parking space? How control the rapacity of the automobile? The GLC still seeks to be on the side of man.

Some seven thousand officials and other men work in County Hall, the headquarters of the GLC, which squats with solid immensity on the south side of the Thames opposite the Houses of Parliament. Westminster Bridge connects the two great organisms. County Hall, built in 1912, has steep black roofs, a green tower, and a crescent-shaped portico on the river; covering no less than 6½ acres, it has 5½ miles of corridors (one is 220 yards long), and rests on a kind of concrete ramp or bulwark to keep the Thames out. An indestructible monolith, it survived thirty-one bombings during the war.

The Greater London Council is a kind of Parliament in miniature, a sister to the national Parliament across the river. Elections held every third year put a hundred councilors and sixteen aldermen into office; the former serve three years, the latter six. The chairman, like Mr. Ferguson, is selected by the Council each year from the body of aldermen, and represents the majority party, either Labour or Conservative, as does the vice chairman; the deputy chairman represents the minority. But during their terms of office none of the three takes part in any political activity whatever; in fact, they are *required* to be nonpolitical. This makes for a healthy nonpartisan approach, and does away with the more revolting types of hurly-burly distinguishing politically run municipalities elsewhere.

The executive is known collectively as the Dais. Below this the aldermen and councilors operate on a party basis, Labour or Conservative, exactly as in the House of Commons, and the majority leader is the London equivalent of the Prime Minister in the national government. As of the time I visited the GLC this personage was Sir William Geoffrey Fiske, one of the ablest officials in the inner circle of British public life; he is sixty-one, six foot five, of large girth, baldish with spines of dark hair, heavily bespectacled, a former official of the Bank of England, and a strong man of the Labour Party. Recently he became chairman of the Decimal Currency Board set up to supervise the impending transformation of sterling into a decimal system. He has a reassuring voice, a fondness for music, a lucid expository mind, and an abundant as well as ironic sense of humor. He sought to explain the London political situation to us, and said that the area of basic agreement between the two parties is considerable, more so than in the national Parliament. But Labour is more attentive to culture, and wants to spend more money on the social services.

Municipal legislation is debated and passed much as national legislation is in the Commons across the river, but normally the Council does not meet more than once a fortnight, and an adverse vote cannot force a Council government to resign, as is the case in the House of Commons. Members serve a full three-year term no matter what. Another striking difference is that councilors and aldermen *are not paid*. Only civil servants and administrators receive salaries. The *elected* representatives work for nothing. This is one reason why London politics have always been so clean. There is absolutely no corruption here if only because the legislators have nothing to gain financially by being elected.

Plenty of men advance from the Council to the Commons—go "across the bridge." Theoretically a man can be both a Councilman and an M.P. at the same time, but few—only one person at present—have the stamina or time for such a demanding double harness.

The GLC has a broad spectrum of functions, from planning on the highest strategic level under the blueprint of the Greater London Development Plan to planting bulbs in 160 parks—a million and a half of these last year. Its annual budget, including education, is £245 million, almost that of a country as big as Brazil, and it is served by more than a hundred thousand employees. It maintains 1,127 schools, supervises the main metropolitan water courses, maintains the Crystal Palace, disposes of 470 million gallons of sewage daily, which is a lot, and censors the movies. Censorship of stage productions is, however, still in the hands of the Lord Chamberlain, an official of the Queen's Household—another example of the extraordinary number of fish in the London fry.[1]

The Borough Structure

Government in London is, clearly, a multiple and complicated process. The two main tiers of authority are the GLC, which we have just inspected, and the boroughs (separate districts). But these by no means complete the administrative picture. The GLC and the boroughs, all with their independently elected officials, have, incredibly enough, no direct power or authority in a variety of vital fields, even including the police.

Consider, too, in passing such huge instrumentalities as the Port of London, which has jurisdiction over ninety miles of Thames-side and operates five huge docking systems, and the London Transport Executive, founded by Herbert Morrison, which handles the movement of millions of citizens daily by bus and underground over 2,000 square miles. Then there are the Metropolitan Water Board (serving 570 square miles), the 235 square miles served by the Post Office, four different electricity boards, and three gas boards. Hospitals are under the jurisdic-

---

[1] The Lord Chamberlain's power of censorship was abolished since this chapter went to press.

tion of four metropolitan Regional Hospital Boards. Parks come in four different categories: those most famous like Hyde Park and Regent's Park are not run by the municipality at all but by the Ministry of Public Buildings and Works, a national cabinet office. The Zoo, one of the finest in the world, is operated by a private society.

To proceed: The boroughs, a series of separate "cities," are the constituent parts of the metropolis, much as Brooklyn and the Bronx are parts of New York City. They became "metropolitanized" by the reorganization of April, 1965, and are now considered to be "primary units" of government, altogether autonomous and not under the jurisdiction of the GLC. The boroughs maintain a staff almost twice as big as that of the GLC—around 200,000 people. Of course they cooperate with the GLC, and liaison is close.

Some boroughs are very large units—Lambeth has 340,000 people, Wandsworth 335,000. Several are fabulously rich, like the "City" of Westminster. The recent reorganization did away with the old administrative units, the names of some of which had been celebrated for generations, and new place names had to be created. Wembley and Willesden became Brent; this was because both Wembley and Willesden felt so strongly that their own names should be perpetuated that Brent, the name of a neighborhood river, had to be chosen as a compromise. Hammersmith is now an amalgam of Fulham and the former Hammersmith, and is nicknamed Full Smith. Newham is a merger of East Ham, West Ham, and parts of Woolwich. The former West Hammers call the new borough "New Ham"; the East Hammers call it "New'em." The celebrated boroughs of Bermondsey, Camberwell, and Southwark have all become Southwark, and Paddington and St. Marylebone have been incorporated into Westminster. Kensington and the Royal Borough of Chelsea steadfastly refused to give up their historic names and consequently the new unit is called "Kensington *and* Chelsea." Bethnal Green, Poplar, and Stepney—famous names all—have been consolidated into a unit called "Tower Hamlets."

Each borough has its own mayor, aldermen, councilors and a town clerk, who is the chief administrative officer. He is a permanent official, appointed by the borough council, and he receives a modest salary. The mayor, like the chairman of the GLC, gets no pay at all; largely a ceremonial official, he serves for only a year. The aldermen and councilors, standing on a party basis, and elected for three-year terms, also work without pay. How much time these public servants devote to their function depends on personal circumstances. They are often dirt-poor, and work for the boroughs in the evenings or in spare time from their jobs—a volunteer effort in the main.

The GLC and the thirty-two boroughs share several important functions, but operate independently in others. In the GLC jurisdiction are main roads, fire prevention, ambulance services, licensing of motor vehicles, flood control, and education within Inner London, as well as planning. The borough councils handle maternity and child welfare, local sanitation, refuse collection, slaughterhouses, traffic, libraries, noise and smoke abatement, and services for the aged, blind, and helpless, amongst much else.

I went to see A. D. Dawtry, town clerk of Westminster, a competent, articulate fifty-one-year-old Yorkshireman, in his offices in Westminster City Hall. This is a large, smart office building on Victoria Street built like glass cages in the most contemporary manner, so new that my taxi driver—a good old experienced London hack—didn't know where it was. Mr. Dawtry, who is also Secretary of the London Boroughs Association, showed me the brilliant view from his eighteenth-floor windows, and talked about the large scope and range of Westminster; it holds Mayfair and Piccadilly, Buckingham Palace and St. James's Palace, the Houses of Parliament and the Abbey, Bayswater and St. John's Wood, Marble Arch, four great railway stations, most of the important embassies, hotels like the Ritz, the theatrical district, and the West End department stores. The borough staff numbers about five thousand. He mentioned traffic as his biggest problem. The authorities could alleviate this by setting up parking space for ten thousand cars in parks and squares, but they will not give up their precious open spaces. In the race between man and the machine, the theory here, as in the GLC, is that man must be served first—if possible.

We climbed to the twentieth floor—it's still a comparatively novel experience in London to rise so high—and waited in the Lord Mayor's "parlour" (an authentic old London touch) to be received by the Lord Mayor. This vigorous personage was A. L. Burton, forty-seven, a Londoner who has been in the furniture business for many years. Again we have an example of the rise of middle-class citizens to exalted positions without regard to class.

The Yard

London is probably the only major capital where the local authorities (except in the City, which has its own force) have no jurisdiction over the police. In most great cities, in fact, the police are the dominant municipal problem. In the United States in particular a municipal administration is apt to stand or fall on its police because of corruption. But this is not so in London, where municipality and police are altogether separated. The Metropolitan Police District, known universally as Scotland Yard, derives its authority from the Home Office, a department of the national government, and is responsible not to any London official but to the Home Secretary, a member of the national cabinet.

The district covers 780 square miles, an area substantially larger than Greater London, and has a personnel of 18,303. Its Criminal Investigation Department operates on a national level as well, in that Scotland Yard is likely to be invited in on any criminal case in the British Isles—also, British dependencies anywhere in the world. It performs various duties such as the protection of eminent personages who in the United States, as an example, are handled by agencies like the Secret Service, and has additional functions resembling those of the FBI and the immigration authorities in the United States. Cooperation with

American agencies and with INTERPOL, the International Criminal Police Organization, is intimate.

I called on Sir Joseph Simpson, the Commissioner, and head of the Metropolitan Police Force, in his office at New Scotland Yard, close to Parliament and the river. This headquarters is a dowdy, uncomfortable building which was once, of all things, an opera house owned originally by the kings of Scotland, from which derives the name Scotland Yard. It is bursting at the seams nowadays, and a splendid new skyscraper has gone up on Victoria Street, a few blocks away, to replace it as the home of the organization. The move took place a few months after my visit. In Westminster City Hall I heard an official mutter, "Scotland Yard's new place is going to cut off our view of the Abbey and Big Ben."

I can't imagine that anybody can walk into the premises of Scotland Yard without a *frisson*. Sir Joseph Simpson sat calmly at a big oval desk smoking a pipe. He was bluff, courteous, and modest, conveying not merely a note of power and application to his job—a cop pure and simple—but of kindness at large and decorum as well. A large man with large hands and wavy silver hair, he stood about six foot three and had a strongly jutting chin and nose. While talking he fiddled with files held together by bits of colored string, in the British manner.

I asked this decent, conscientious man about his career, and he told me that, born in Shropshire, he joined the police force there in 1931, when he was twenty-one. He had been a policeman ever since, climbing up steadily from the humble rank of probationary constable, and is, with one exception, the first career officer in the Yard's history to rise to be Commissioner. One of his sons is a farmer, one a policeman in Hong Kong. His salary is £8,600 a year.[2]

The Metropolitan Police District operates on a budget of roughly £45 million a year, half of which is paid by the Exchequer, half by local taxes. The functions of the police are broad in

---

[2] Sir Joseph Simpson died after these lines were written. His successor, Sir John Waldron, had been his deputy since 1966.

England, and every officer is held to be not merely a servant of the police authority, but an independent holder of a public office himself. Corruption is unknown. In eight and a half years as Commissioner, Simpson had to deal with only one minor case of bribery. The crime rate is going up, but it does not reach anything like the bewildering level it has in several American cities. The murder rate in London has held steady at about 35 a year for a decade, although it rose to 49 last year. In New York there were 637 murders in 1964, 653 in 1966. But indictable crimes as a whole have increased three and a half times in London in the last few years, and robbery with violence, which Sir Joseph called "the crime of the day," has gone up 450 percent. "Violence" does not, however, necessarily mean armed violence under British law, and may merely indicate use of the fists, kicking, or the like. The technical term is "theft with force." Even so, its increase is worrying.

Drugs are a mounting problem as well. A good many youngsters smoke marijuana or take LSD, but, since these drugs are held not to be habit-forming, they are of comparatively small concern to the police. Heroin is another matter. It is dealt with in London in a novel manner, which sometimes startles authorities in America or on the continent. Addicts are perfectly free to buy heroin legally on a doctor's prescription. This, in the British view, tends to discourage traffic on the black market, which is what leads to crime. If an addict is free to buy an addictive drug legally, he doesn't have to rob or otherwise act illegally to support the habit. A few addicts may always be seen outside one well-known all-night pharmacy in the West End waiting for 12 P.M., when they can buy the next day's dose. But there are only about a hundred known addicts in England, as against more than sixty thousand in the United States; those in London are familiar to the police and are watched carefully, but they are not considered to be criminals. "Any form of prohibition is apt to do more harm than good," Simpson told me. Yet he freely admitted that an increase in heroin addiction in the youth recently was disturbing, although it comes nowhere near the level of New York.[3]

London bobbies, as is well known, do not carry arms except in special emergency circumstances, and this has provoked a bitter controversy since the wanton murder of three unarmed policemen by thugs in a London street last year. A public outcry demanded that the police should at least have a chance to defend themselves. Scotland Yard authorities in general seem to feel that the old tradition against carrying firearms should be maintained for two reasons: first, if the police carried guns, it would encourage criminals to do the same; second, police power should rest on the basis of complete friendly cooperation with the people, which is aided by having a nonarmed force.

At Last the City

London's central bastion, the heart of all this mastodonic complex, is "the City." A tiny enclave covering 1.03 square miles, it lies almost in the exact center of the megalopolis, and is the financial nexus of the realm. The population is around half a million by day, but less than five thousand by night; few sleep here. This is the abode of business, business, business, and, like Wall Street, it is almost deserted after office hours.

Officially the City has no connection at all with the rest of London, strange as this may seem. A corporate entity of its own, it is operated autonomously by what is known as the "Corporation of the City of London," and is quite independent of the Greater London Council and the boroughs, although it cooperates with them. Its celebrated Lord Mayor, who is often erroneously thought to be chief executive of all London, has no jurisdiction whatever outside the City's limits. The City even has its own police force, independent of Scotland Yard. The City police wear

[3] Since this chapter was written there has come a change. A bill has been introduced in the Commons to control narcotics prescriptions, and several delinquents have been arrested and jailed for "the smoking of Indian hemp" (marijuana).

Roman helmets, red and white armlets, and uniforms heavy with brass buttons; every man must be five foot eleven or taller, and the average height is six feet. Its members carry a kind of radio speaker for quick communication.

For several centuries the City was the most conspicuous agglutination of financial power in the world. It contains the Bank of England (nationalized in 1946), Lloyd's, and the Stock Exchange. More foreign banks are represented here than in any city in the world, as well as no fewer than four thousand British banks. It does a prodigious business in insurance, and it finances Britain's export trade, which is still worth around £400 million a year in spite of contemporary troubles and harassments; its invisible earnings alone have reached £150 million in a single year.

Much else is located within this unusual 667 acres. The City operates four London bridges, includes long miles of docks, contains both Fleet Street and St. Paul's, and owns several of the great markets like Billingsgate, Spitalfields, and Smithfield, which feed the metropolis. Old Bailey, the central criminal court, stands here on the site of Newgate Prison. The City even owns far-off open spaces like Epping Forest in Essex and Burnham Beeches in Buckinghamshire, the maintenance of which comes from "City's Cash," or private purse. What really counts, however, is the City's control of finance, mercantilism, trade, gold. Its essence is the power of money.

Origins of the City, where messengers still wear top hats, go all the way back to the Romans, and its present boundaries were fixed, I was told, in Norman times. Much of British history for centuries hinged on a kind of unspoken struggle for power between the moneybags of the City and the Kings and Parliament at Westminster, because rule without money was impossible. The merchants held the strings. Right up till today the British monarch must ask the Lord Mayor's permission to enter the City. The Queen, as of the moment, and her entourage present themselves at Temple Bar on the Strand, the City's frontier, where the Lord Mayor greets them. (The actual bar, built by Christopher Wren, has disappeared, but the line of demarcation is still the same.) The Lord Mayor, in a gesture of obeisance, presents to the Queen one of the four historic swords of the City, with its tip pointed downward; duly she hands this back, and he carries it away erect as a symbol of his authority. Honor is served both ways. The Lord Mayor kowtows, then rises; but the point is clear that the Monarch must ask permission every time she enters the City's august preserves.

The City is governed in a somewhat complicated manner. Commerce and panoply are interwoven. This is a plutocracy like few surviving in the modern world. The heart of the structure is what is known as the "Common Hall," which is made up of representatives of the "livery companies." There are eighty-four of these, each being a kind of guild descending from medieval times. They arose originally to protect the interests of their members, each following a craft, and gained prestige and commercial power through centuries. Several have halls where ceremonial banquets of the most dazzling splendor are still given once a year. But few of the livery companies perform actual civic functions any longer, or are even connected with their original business; a man can be invited to join the Haberdashers, as an example, without being a haberdasher. The Goldsmiths do, however, still put their hallmark on metal, and the fishmongers regulate some aspects of the fishing trade.

Pages could be written about the picturesqueness of the livery companies. Among the twelve so-called "great" companies are the Mercers, founded in 1393, the Skinners (1327), and the Salters (1558). The oldest is the Weavers (1184). Among the newest are the Air Pilots and Navigators (1955), the Tobacco Pipemakers and Tobacco Blenders (1960), and the Scientific Instrument Makers (1964). Others are the Armourers and Brasiers, Bowyers, Cordwainers, Lorimers, Scriveners, Gold and Silver Wire Drawers, Saddlers (1272), Broderers, Girdlers, Fletchers, and Apothecaries (1606). This last still has an examining body which grants diplomas in "medicine, surgery, and midwifery."

Several livery companies have become extinct, like the Silk-Throwers, Hatband Makers, and Long-Bow Stringmakers. The biggest is the Shipwrights, with five hundred members; the smallest is the Ironmongers, with thirty-five. Two guilds, the Parish Clerks and Watermen, have never been granted a livery, which means that they are not "in." Thirty-three companies have halls, and there are about fourteen thousand liverymen in all; at official occasions their representatives wear robes or uniforms which make the costumes of Mods or Rockers look demure.

The City electorate is small, since almost all its population lives outside. To vote, a man or woman must be an actual City resident or the owner or tenant of City property with an annual gross revenue exceeding £10. All candidates for City office, who must be liverymen, stand as individuals, not as representatives of the political parties (though, of course, most important figures in the City are Conservatives). No partisan politics are involved. Twenty-six aldermen are chosen—one for each of the 25 wards, with an extra one thrown in—and 159 councilors. The aldermen serve for life, the councilors for a year. Some wards bear pleasantly outlandish names, like Cheap, Farringdon Within, and Farringdon Without. Two sheriffs (this office dates from the seventh century) are also elected, but in a different manner—directly by the liverymen sitting in Common Hall.

At the top is the Lord Mayor, who changes every year. There are always two candidates for this office nominated by the liverymen by a show of hands; each must be an alderman who has served as a sheriff. The procedure is pretty much cut and dried; gentlemanliness rules; everybody knows who the two candidates will be, and the loser will probably become Lord Mayor the next year. The actual choice is made in the Court of Aldermen by sealed vote. When the Lord Mayor gives way to his successor, the ceremony is known as the "Silent Change." After serving his year's term he resumes duty as an alderman, and is known as one "who has passed the civic chair." The incumbent is obliged, by immutable tradition, to turn over to his successor a crystal scepter "used at coronations," as well as a "tattered purse" given to the City by Queen Elizabeth I.[4]

[4] *New York Times*, November 11, 1967.

A fantastic amount of pageantry is attached to all this ritual of government. Before the ceremony of inducting a new Lord Mayor the principal City officials, in scarlet robes, attend service at St. Lawrence Jewry, a Wren church; this tradition began with Dick Whittington, who was "thrice Lord Mayor" between 1397 and 1423. The procession returns to the Guildhall, built in the fifteenth century and badly blitzed in the twentieth, and approaches the Hustings, a kind of platform which is strewn with scented sweet herbs—a "precaution" against contagion in earlier days. "All in the procession," says one account, "carry little posies of old English flowers."

Later comes the Lord Mayor's Procession, or Show, one of the grand old sights of London. Its ostensible purpose is to enable the Crown, which is represented by the Lord Chief Justice, to give approval to the election; another is to amuse and impress the populace. The Mayor rides down the Strand to the Law Courts in an ornate gilt coach which, built in 1757, weighs nearly four tons and is drawn by six stupendous horses. Pikemen in uniform guard him, and the scene shimmers in scarlet, maroon, and gold. Following come a series of floats with colorful displays, some of them vulgar. The apogee is the Lord Mayor's Banquet at Guildhall a few days later, which is the single most ceremonial meal served in England; the Prime Minister of the day makes the principal speech, and all the great dignitaries of the realm are present, costumed to the nines.

The prestige of the Lord Mayor is, indeed, high, although his power is almost nil. The paradox is clear that, although he has no authority in London itself outside the City, he is a national rather than a municipal figure. Within the precincts of the City he takes precedence over everybody except the monarch, even over other members of the Royal Family. The Lord Mayor gets no salary as such, but his representation allowance is £15,000 for his year of office, a tidy sum. Even so, most are obliged to spend a good deal of their own money in order to cope with the cost of various entertainments. No poor man ever dreams of becoming Lord Mayor, and, conversely, nobody ever seeks to run for alderman unless he is aiming to be Lord Mayor some day—and can afford it.

The Lord Mayor is Chief Magistrate of the City, and a trustee (as the quaint language goes) of the "Fabric" of St. Paul's Cathedral. Four times a year the passwords of Buckingham Palace and the Tower of London are delivered to him "under the sovereign's signed manual," although neither the palace nor the Tower is located within the City. Assisting him in his duties are dignitaries with such titles as Secondary, Remembrancer, Sword Bearer, and Common Cryer. He must be resolutely nonpolitical during his year's term of office, in the familiar London manner, and be prepared to make at least a speech a day as well as innumerable other public appearances. Meanwhile the actual, down-to-earth administration is done by the Town Clerk.

The Lord Mayor when I last visited the City was a fifty-six-year-old merchant, Sir Robert Ian Bellinger, alderman from Cheap and a member of two livery companies. He told me proudly that he was the 639th Lord Mayor of London in a continuous sequence since the twelfth century. A self-made man, Sir Robert had little formal schooling after the age of fourteen. His father was a catering manager, who died when he was eight. He got a job as an office boy, and started up the ladder at fifteen shillings a week, the equivalent of $3.75 at that time. Sir Robert has sharp eyes and a narrow, decisive face, with dark-reddish eyebrows and thin lips. His drive and substance are clearly apparent. He is fast-pacing, fast-speaking, and at times his voice lowers to a hard, confidential whisper. He says "eyed" for aid and "pied" for paid. Again we have the phenomenon of a man reaching great station in contemporary London without the conventional advantages of wealth, education, or social class. The old forms are breaking up.

Sir Robert received us in Mansion House, the Lord Mayor's residence, which was built in the 1740's on the sight of the old "Stocks" market. He talked about national finance and the necessity for strict economies, and then took us for a swift tour of

Mansion House. This is the only private dwelling in the world containing a functioning court of law, called "the Justice Room." Offenders picked up for various petty crimes and derelictions in the City are held here and tried. Those accused rise from cells underneath the floorboards by a short flight of stairs, and the dock is small. The Lord Mayor adjudicates on cases twice a week, and is the only lay magistrate in England who, sitting alone, is empowered to give sentences up to six months, although he is neither judge nor lawyer. To the wonders of the British legal system there is no end.

The City is probably the repository of more antique tradition and fixed institutional force than any square mile on earth, but in several respects it is changing just as the rest of London is. The skyline differs beyond belief from what it was before the war. The Blitz destroyed more than a third of the City's floor space; whole square blocks were wiped out, including eighteen livery halls and twenty churches. Reconstruction is going on smartly—much of it in the modern manner, with tall glass buildings—and the area will, it is hoped, be fully rebuilt by 1980. One innovation will be the use of "pedways" so that pedestrians can traverse various sections without having to pass through traffic. A device known as the "travolator" is already functioning—a moving platform, the first in Europe, to expedite movement in one of the big underground stations. Moreover, the Barbican development near Mansion House is turning a blitzed district covering sixty-three acres into a residential or community area designed to encourage people to live in the City, not just work there—a city within the City. It will be built on two tiers to alleviate traffic. A students' hostel, tower apartments rising forty-four floors, other large housing developments, a new home for the Royal Shakespeare theater, an elevated walkway system, facilities like swimming pools and tennis courts, art and shopping centers, are included in this ambitious project. Even in the City, London is on the move, and the planning concept is being actively applied under an antique plutocracy.

Vistas Across the Belt

The most challenging of all developments in changing London are the new towns, which embody a daring experiment—that of providing for the "overspill" of the metropolis by creating new communities outside. No innovation on quite this scale has ever been attempted by a metropolitan administration before, although the move outward resembles to a degree the trend in American cities for businesses to establish themselves in the suburbs.

As to London, what the authorities want is to stabilize Greater London at its present population of eight million plus. The numbers of people moving in and leaving cancel each other out, so that the core of the problem is to deal with the city's natural growth. The present population will, demographers say, generate another million by the 1980's. Hence to achieve stability near the eight-million mark about a million people will have to be "exported" in the next twelve years. But where? The impulse is to build out, not up. Yet the metropolis is, we have seen, firmly encircled by its precious and inviolable Green Belt, which must not be tampered with or built upon. So the bold decision was taken to leap right over the Green Belt and construct new cities on its other side.

These are not intended to be commuter cities—indeed, the opposite is true. They are not "bedroom towns," like Rye or Port Chester near New York. The theory rested on the principle of keeping people out of London, not of sucking them in. The new towns are supposed to be an answer to the commuting problem, with all its gross burdens, not an adjunct to it, and are designed to meet the goals of industrial planning as well. Indeed, a basic function of the new towns is to house new light industries outside London and thus stimulate industry at large, encourage new manufacturing enterprises, and transfer people from the overburdened service industries in London itself to productive industries outside.

There have been difficulties aplenty. Many Londoners, particularly skilled workers, don't want to move. Then, too, much of southern England has already become covered by a web of comfortably fixed light industry, which makes it more difficult to entice citizens into the new experimental towns.

The New Towns Act passed by Parliament in 1946, under a Labour government, initiated this vast experiment in urban management. Eight new towns have been built so far—like Basildon, Harlow, Hemel Hempstead—with others in the planning stage. They are run by government-sponsored corporations, and are financed by the national Exchequer. Additionally, the GLC and local authorities are working jointly on the expansion and modernization of several old existing towns, such as Ipswich, Peterborough, and Northampton. The aim is not merely to assist the movement of population but to make room for urgently needed developments in the reconstruction of London itself, which overpopulation inhibits or impedes.

The old "generation" of new towns, built in the 1950's or thereabouts, have populations of 60,000 to 80,000; those newer are being planned to hold 150,000. All are made on the same pattern. They look "modernistic," with standardized (but moderately attractive) housing, effective techniques in traffic control, bright rows of glass-bound factories, carefully fashioned open spaces, shopping centers, and alert new schools. Perhaps the atmosphere is somewhat aseptic, a bit arid and forbidding. Some new residents, accustomed to the hurly-burly of the giant metropolis, with its illimitable profusion of attractions, its spread and sprawl and sparkle, feel like pioneers in a cold, mechanical wilderness—lost, sterile. The atmosphere is almost that of science fiction or Brasília.

I spent a revealing afternoon at Stevenage, Hertfordshire, a new town thirty-one miles from Charing Cross on A-1, the Great North Road, population 58,000 and growing by about 3,000 people a year. Stevenage (the name means "Strong Oak") *was* an old town—a village of some 6,000 with a church dating from the twelfth century, and a powerful wing of its citizenry opposed vigorously the idea of re-creating it into a new urban laboratory.

But the dissidents have become largely mollified by now, if only because the experiment has brought prosperity to a degree.

The new Stevenage is a good example of contemporary design. Lorry traffic is routed around the town, so that noise and petrol fumes are eliminated. There are separate cycle and pedestrian paths as well as roads; a person can walk or bicycle through all the town's fifteen square miles without encountering any heavy traffic. The maximum distance anybody has to go in order to get from home to work is two and a half miles. Fifty industries are represented in the brightest, shiniest of new plants—"small-unit" factories covering a total of 3½ million square feet. Kodak has an installation here, and so do the British Aircraft Corporation (4,500 employees), Hawker-Siddeley, and International Computers. Stevenage makes electronic equipment, guided missiles, conveyor belts, rocket gadgetry, tabulators, and a variety of other technical products that symbolize the new industrial and scientific age.

The community, I found, is organized tightly into six neighborhoods with populations of about ten thousand each. Each has three shopping centers, one big and two small, which are identical in each neighborhood. The small centers have six different types of shop side by side, neat as needles, which make an interesting cross-section of what the average Briton considers essential —a hardware shop, a tobacconist (who also sells newspapers and confectionery), a hairdresser, a butcher, a greengrocer, a provisions merchant, and a pub. The larger centers—one for each neighborhood—have forty-two other types of enterprise, like banks, dry cleaners, a furniture store, a restaurant, and so on. Everybody has TV, but there are no aerials on individual houses; a master system pipes the channels in. But there is only one movie—for fifty thousand people!

The public corporation running Stevenage, which has cost £45 million so far, has erected thirteen thousand new "system-built" dwellings, most of them two-story houses with three bedrooms, a bath, a toilet, and a minuscule garden. Deliveries of goods take place in the rear of the house, not the front. There are "flatlets" for old people, a library, facilities for sport, various public services, a workshop for the physically handicapped, a cooperative wine society, and three weekly newspapers.

Previously I had visited some of the dilapidated areas in London from which most residents of the new towns come. They were intolerably dingy, sordid, down-at-the-heel. Here, in sharp contrast, replacing squalor, are brightness, fresh air, educational opportunity, good jobs, and the stimulus of being part of a model "frontier" community, even if the atmosphere resembles *Brave New World*. A week's rental in London for a miserable single room can be the equivalent of $30 a week; here a whole neat house (unfurnished) costs $19.40. (One reason why rents are so high in several districts of London is the growth of new black-white districts; landlords raise rents in an attempt to keep Negroes out.) Of course it should be reiterated that renewal projects are also being pushed in London itself by the Greater London Council and the boroughs.

A newcomer to Stevenage or the other new towns has to satisfy only two conditions—come from London and have a job. The community does not screen applicants, but the employer does, after setting up his factory. What the new towns want most are skilled industrial workers and experienced technicians. The average family income is around £20 a week.

Exit

I have used a good many comfortable-seeming facts and figures in this text, which perhaps reflect unduly the euphoria that distinguishes much of London today. So a realistic warning should be added. There are wretchedly poor as well as rich people here. London is the capital of a country gripped by a widespread, deeply based, seemingly permanent, and perhaps ultimately insoluble economic crisis. The mood is up, yes, and illimitable sums are spent on such an irrelevance as gambling, but it should not be forgotten that this country cannot continue to survive unless the trade balance is substantially improved and domestic production built up, two things difficult in the extreme to do at the same time.

Similarly, the prevalent euphoria among youngsters has its deleterious aspects—frivolousness, sloth, insularity, waste. Caste is still a blight, even if strictures imposed by the old-school tie are loosening. The conduct of public affairs often seems to lack grip, cohesion, direction, and the will to win.

As to London itself and its piebald assembly of different regimes and governments, the miracle is that it all works so smoothly. The Crown, the Parliament, the national government, the GLC, the boroughs, the City, the public authorities, the police, the Post Office, the new towns, with all their complex overbearings and underlappings, their seemingly fortuitous and capricious mixture of forms, their contradictions, inefficiencies, and medievalisms, somehow combine to make a smooth and civilized amalgam. Perhaps this is because the welfare of the citizen is a prime desideratum on the part of almost everyone. The antique becomes the new, but the basic standards do not change. London is like a person who has performed the extreme miracle of getting over old age.

3. IS PARIS DEAD?

Paris blew up early in May, 1968, and the echoes of this convulsive event are still being heard. It has been difficult for disinterested observers to get an altogether coherent account of what happened during several vivid weeks of turmoil, because eyewitness stories varied to a marked degree. The main elements are, however, clear. Students in the Sorbonne and at Nanterre, its auxiliary institution in the suburbs, suddenly broke loose. Rebellious youngsters stormed through the streets in the Latin Quarter and elsewhere—cars were burned, pavements torn up, and windows smashed. There was, however, comparatively little bloodshed; the police did not use weapons other than clubs and tear gas, and only one person was killed. Even so, parts of Paris became a shambles. A general strike followed the student outburst, and France itself was virtually paralyzed, with some 500,000 workers involved. Factories closed down, the dockyards stopped working, and the airports went dead. Newspaper deliveries stopped in Paris, and celebrated restaurants closed their doors. There was no gasoline for motorcars, and countless thousands of Parisians had to walk to work—if any work was going on. "It was like a white death," one Paris friend moaned to me.

Oddly enough, the first outbreaks were, it seems, provoked and led by students on the *right*, not the left. Then the conflagration spread. Universities all over France had to close their doors. Almost at once the insurrection was, in a manner of speaking, taken over by the CGT, the General Confederation of Workers, which is the leading labor organization in the country and which is

Communist-controlled. At least three different varieties of Communist or extreme left-wing elements became conspicuous. One wing of students was strongly committed to Chinese influence—particularly a group known as the Union of Marxist-Leninist Youth. Other wings were nihilist anarchist, Castroite, or Trotskyite. Some had the motto "It is forbidden to forbid." Daniel Cohn-Bendit, a twenty-year-old stateless youth of German extraction, with bright red hair and a personality to match, was the principal student leader.

What led so many students to defiance and open revolt? What stirred them? What made them willing to challenge authority by force and face tear gas and fierce clubbings in the streets? First, an altogether genuine and justified resentment at the inadequacies, apathies, and archaic structure and functioning of the French university system. Second, hostility to the lunar figure of General Charles de Gaulle—also sheer boredom with him, as he prepared to enter his eleventh year in the presidency. Then, too, such matters as youth versus age, restlessness versus conformity, a desire for improvisation and experiment as against bondage to ossified older forms, played a signal role, just as they have in other recent student demonstrations all over the world.

The government, under the able leadership of Prime Minister Georges Pompidou, did what it could to restore order. This was not enough, and it became necessary to ask General de Gaulle (who was absent on a state visit to Rumania) to return to Paris. Not long thereafter, following a broadcast from this remarkable man, who had obtained the support of the army, the turbulence died down. Elections followed, and, to the surprise of many, the Gaullist party with its retainers and allies won 294 seats in a chamber of 437, giving it an outright majority—the largest it had ever had. The popular vote in de Gaulle's favor was less overwhelming, but substantial. The Communists fell from 73 seats to 34, and the Federation of the Democratic Socialist left, the principal opposition to de Gaulle, lost half its members in the chamber. So once more the haughty seventy-seven-year-old general was confirmed in power and leadership. Presently he tossed Pompidou out of office like a gnawed bone, and Maurice Couve de Murville became Prime Minister. This is not to say that Pompidou is necessarily finished.

✿

But this chapter is not about French politics; it is about Paris the city. It is time to move along. For all its vicissitudes Paris is still all of a piece. It is not a fixed combination of communities, like London, or a collection of scattered villages, like Los Angeles, or a series of pincushions, like New York. Neighborhoods in Paris may differ, but the essential Paris, that of the twenty arrondissements within the walls ("Paris Intra Muros"), is the same organism from top to bottom, from solid rim to rim. You could lift it whole from a plate, like a nicely fried egg.

Not without reason is Paris still called the City of Light, despite the savagery of the 1968 disorders. The grandeur of the old French kings, centering on Paris and Versailles, surpassed that in any other court, and French was a language that every civilized person had to learn. France produced the most superior intellects, the most refined and articulate men of art, letters, and ideas, all the way from its earliest history through Proust. Paris gave joy; it was the most thrilling of all cities in our youth, with its incomparable beauty, magnificence of design, and free creative atmosphere. It was a city which understood the delights women bring; there were few puritan rigidities. Even in the humblest quarters people carried themselves with pride and individuality. And it was marked by a sense of abundant freedom; citizens could do as they pleased—within reason—and nobody got into trouble for being unconventional. A civilized life was possible on almost any level. The burst-out in the spring of 1968 shook several of these conceptions, but the roots and stem of Paris are little changed.

Several distinguishing Parisian characteristics might be mentioned:

Paris expresses well what Gibbon once called the classic virtues—balance, lucidity, precision.

It has always had a gift for drawing creative talent from all over the world, quite aside from what it produced itself. It has a compelling magnetism for temporary visitors or permanent expatriots devoted to the arts. (This characteristic is, however, less marked today than in former years.)

It gives out an atmosphere of superiority, even arrogance. A true Parisian feels that he is something special, faultless, on the summit.

Its residents have a highly sophisticated appreciation of some of the better things in life—quite outside the intellectual sphere—good beds, sheets, food, wine.

It has a sound bourgeois respect for money—in the bank, pocketbook, or mattress.

For all its universality and hospitality to world culture, it has strong xenophobic tendencies.

It has a martial tradition which, at first glance, may seem to contradict the concept that the greatest of all French contributions to society is its *mission civilisatrice*. I don't know any city quite so full of military monuments—the Arc de Triomphe, the Invalides, the Pantheon. An astounding number of great streets are named for military men, like the Avenue Foch. The outer boulevards are named one by one for Napoleon's marshals—Kléber, Masséna, Ney, Murat, et cetera.

Then, too, Paris is France. Berlin is certainly not Germany, not even West Germany, and Rome is certainly not Italy. But Paris *is* France, although large efforts at decentralization are under way.

What Ails Paris?

All this being said, Paris is not what it used to be. The atmosphere is generally down-beat. In some respects this most brilliant of cities seems to be only half alive. This helped to make the 1968 riots seem even more spectacular than they were, because they were so unexpected. Take such a homely matter as building, in which Paris has been atrophied for many years. Hotel space is not a particularly vital criterion, but the fact is remarkable that only *two* new luxury hotels have been built here since the end of World War I—the George V in 1928, the Hilton in 1966. In this respect Paris has become a backwater like Buenos Aires.

What is more, other new building is at the minimum. One reason for this is clear—the commendable passion to keep Paris beautiful. A major constituent of beauty is, in French eyes, harmony, and for many years, in order to preserve the present glorious—but low—skyline, it has been forbidden to build any structure higher than eight or nine stories. Of course this prohibition discouraged private builders, and construction fell off in consequence. Now the taboo on higher buildings is to be lifted, in certain sectors, and change will come. As to public building, it is another astonishing fact that no more than four—exactly four —major new public or semipublic buildings have been constructed in the last twenty years. Among these are the magnificent new UNESCO headquarters; the Maison de la Radio on the Quai Président Kennedy, in which French radio and TV are concentrated; and Maine-Montparnasse, a huge white office building—it houses Air France amongst much else—which dominates a whole neighborhood with a stunning effect.

These are virtually the only new buildings worth mention. Compare this to the bewildering profusion of new construction in New York, Moscow, São Paulo, Brussels, or even London!

Thirty-three bridges cross the Seine, which rolls through the city in three sluggish loops. Only one new bridge, the Garigliano on the far west side, has been built in thirty years. Many of the sewers, gas lines, electrical conduits are antiquated. It takes several months to get a telephone. Paris drinking water leaves much to be desired, which is natural inasmuch as most of the piping has not been renovated since 1895. On the other hand, as we shall see, substantial work is in progress or is being planned for amelioration of traffic on the boulevards, extension of the admirable Métro, the use of tunnels and expressways along the quais, and the creation of new urban work-and-housing projects in the outskirts of the city, such as that at the Rond Pont de la Défense near Nanterre.

Even so, contemporary Paris varies hardly a jot or tittle from the city laid out in all its precise regularity by the formidable Baron Haussmann, its prefect for seventeen years in the middle of the last century. Haussmann, under Napoleon III, worked out plans for the Place de l'Opéra and the Étoile, located the railway stations on the rim of the inner city, and invented the system of diagonal boulevards that have been characteristic of Paris ever since.[1] The Pont Alexandre and the Grand and Petit Palaces came later, carrying on the Haussmann concept, and the Palais de Chaillot was built to celebrate the great Paris Exposition of 1937. Since then there has been little construction except what I have named.

Turn now to other elements in the present Parisian scene, which also seem to indicate sterility or apathy in spite of current effervescence by the youthful disenchanted. The theater is senile —it has nothing of the vitality of, for example, the London theater. The *nouvelle vague* in the movies has not quite lived up to its promise, and the local TV (only two channels) is miserable. The city has only one really first-rate newspaper, *Le Monde*. Forty years ago six or seven of the greatest painters in the world —the School of Paris—lived here; they have passed on, and have not been replaced. The city boiled with artistic and literary ferment in the past generation—cubism, Dada, surrealism, existentialism were all born here—but there are few comparable developments today. Nor is Paris any longer an inevitable magnet and haven for new genius from overseas. One of the great Parisian qualities was permissiveness. Many first-rate writers and artists still live here and take advantage of this—Ionesco, Beckett—but the present colony of expatriates cannot compare with that of

[1] I have heard that Haussmann laid out the long diagonals with their systematic cross points and intersections so that it would be easier for the police, if necessary, to shoot down revolutionary mobs. Also, Haussmann carefully sought to segregate Paris into "rich" and "poor" districts.

the 1920's and 1930's, in which shone such blazing and durable stars as Hemingway, Fitzgerald, James Joyce, Picasso, Stravinsky, and Diaghilev.

Paris has always been a "gay" city, and its night life gave it a quality unmatched by any metropolis in the world. There are changes now. Montmartre is dead, and the girls are off the streets —at least officially. There are still vestiges of animation at the Café Flore on the Boulevard St.-Germain, and it will be hard to find a table at the Deux Magots on a mild spring evening, but such classic cafés as the Dôme, the Rotonde, and the Select in Montparnasse, where giants roamed in the 1920's, seem blighted, even deserted. Nor are the supreme classic restaurants as good as they once were—several, like Larue, have disappeared. It was a shock to discover recently that all three bars at the Ritz shut down at 10 P.M.

To explain all this is somewhat difficult. The blood drain of two world wars probably has something to do with it, together with the humiliation of defeat and occupation by the Nazis in World War II. Another is the imposition of drastic taxation in the postwar period, rising costs, and fantastically high prices. Moreover, today, taxes are actually collected, and Paris is, as everybody knows, probably the most expensive city in the world after Caracas and New York. Twenty-five years ago people didn't have to earn so much to keep their place in society, and lived under a good deal less financial strain. Then, too, we must mention again the significant person of General Charles de Gaulle. He suggests—in fact, imposes—an atmosphere of restraint, conformity, morality, even of priggishness. This is particularly hard, among other things, on intellectual life and the arts. The essence of Paris used to be freedom, as I have mentioned above. It was freedom above everything which gave the city its extraordinary exuberance and vivacity, but under de Gaulle—however large and significant his political contributions may be—a good deal of creative spirit and individuality has been lost.

In any case, the contemporary Parisian mood—of course these generalizations may be too broad—includes nervousness, depression, obsolescence, and fear of isolation. People worry that such adventures as de Gaulle's visit to Canada in 1967 may make the nation appear ridiculous. One of these days, they say balefully, he will get so far out on a limb that he can't get back. Moreover, nobody knows what will happen after his death or, as is conceivable, if he should be turned out of office in some forthcoming election. There is no discernible mechanism for succession.

One French friend of mine puts forward another reason for the great change in Paris—the automobile, because it has killed courtesy. Still another is technological advance. Most French people do not like industrialization in all its aspects, and do not welcome newness for its own sake. Of course they use computers in factories and electric dishwashers in the home, but many still prefer the older way and the gentler pace.

The Governmental Structure

Technically speaking, there are three different organisms today all of which bear the name "Paris," following a complex reorganization of the city dating from the early 1960's. Before this there had been no substantial change since Haussmann in 1858; the basic administrative charter still dates from Napoleon.

First, *Paris Intra Muros,* that is, Paris proper, the nut within the shell. Its frontiers, the Boulevards Militaires, are intersected by twenty-one portes, or gates—like the Porte Maillot—and it is divided into the twenty administrative units known as arrondissements, each with its own mayor. The area of Paris Intra Muros is 33.5 square miles (excluding the two great parks, the Bois de Boulogne and the Bois de Vincennes), and it is thus slightly bigger than the island of Manhattan (31.2 square miles). The population is around 2.8 million.

Second, Greater Paris or, as it is known officially, the *Paris Urban Complex.* The area of this is 463 square miles (a bit less than half of Rhode Island) and the population 7.4 million. It is the most densely populated metropolitan area in the world after

Tokyo. It includes various well-known suburban areas like Neuilly.

Third, the *District of the Region of Paris,* which covers a much wider area—4,670 square miles, twice the size of Delaware—and has roughly 8.5 million people, or 17 percent of the population of all France. It contains the whole of three great former departments, Seine, Seine-et-Oise, and Seine-et-Marne, which have been broken down into eight new administrative units, as well as such communities as Versailles, Fontainebleau, St.-Germain-en-Laye, and Rambouillet.

Politics played little role in this reorganization, which was put into force largely to promote an over-all integrated planning scheme. But also it follows tradition in that it has been axiomatic ever since Louis XIII that rule in Paris should always be divided. This is because the national government—of *France*—has always distrusted the Paris "mob"; the great city must never be permitted to get out of hand, since he who rules Paris rules the country. Division of power in Paris has been an obsession since the revolution.

Consequently, and for reasons of rationality and convenience as well, three different principal authorities exist in Paris proper today, which is officially known as a "public collectivity"—the *Conseil Municipal,* the Prefect of the Seine, and the Chief of Police.

The Municipal Council has forty members, who serve six-year terms and who are elected by the people. It meets regularly in public sessions to "propose, discuss, and vote" on such matters as housing, schools, the fire department, hospitals, and the like. But the vote of this Council means comparatively little, since either the Prefect of the Seine or the Chief of Police, both of whom are officials appointed by the *national* government, must approve of its proposals before they can be implemented. A further authority lies in the mayors of the twenty arrondissements, who have substantial local power. Each arrondissement constitutes a separate judicial district, with its own "Justice d'Instance" (justice of the peace), who can be an important personage in France. The twenty mayors serve out of ambition or public spirit, since it is a rule that they receive no salary—like their counterparts in London.

The Municipal Council elects a President out of its own membership each year, who becomes the "Mayor" of the city, for a twelve-month term and whose duties, like those of the Lord Mayor of London, are largely ceremonial. The present Mayor, who took office last year, is Michel Caldagues, a lawyer and pronounced Gaullist. He is only forty—the second youngest man in Parisian history ever to become President of the Council. The youngest was Georges Clemenceau, who reached the post at thirty-six. The only Mayor ever to be re-elected several times was, oddly enough, Pierre de Gaulle, brother of the General, who served from 1947 to 1951.

I had long and instructive interviews with the three men who count—Maurice Doublet, the Prefect of the Seine, Maurice Grimaud, the Chief of Police, and Paul Delouvrier, the Prefect of the District of the Region of Paris.

A word on each:

M. Doublet has his office in the Hôtel de Ville, or City Hall—a large, old-fashioned room overlooking a bustling view. A *huissier* dressed in the most formal fashion received us; he wore white tie and tails, like a nabob going to a ball, and had around his neck, as is typical of guardians of the gate to the offices of French dignitaries, a silver chain resembling that which wine waiters wear. Doublet himself wore ordinary civilian dress. He is fifty-three—a veteran civil servant—appointed directly to his post by the President of the Republic, with a staff of 43,000 and a budget running to the equivalent of $667 million a year. De Gaulle promoted him to Paris from Grenoble, where he had also been prefect, in August, 1966. Short in stature, pleasantly stout, with white hair *en brosse,* he has an alert and cheerful manner. Doublet was born in Bordeaux, the son of an officer. None of the men who run Paris today are Parisians. He told us that this is an advantage—it gives these officials a better perspective, a longer view.

The Prefect executes the decisions of the Municipal Council, and is the link between this body and the national government. A modest man devoted to his job, Doublet is rarely seen in public. He works twelve, fourteen, sixteen hours a day, as one of the "new men" creating the "new Paris." He threw up his hands—jovially—when we asked him what his chief problems were. He handles matters all the way from street cleaning to traffic lights, from control of the bridges to municipal housing.

Paris is changing fast, he thinks. There is much less segregation between the wealthy and the poor than in former days, because a general leveling off is taking place. The great banks have been nationalized, as are the railways, Air France, TV, the radio, and several large industries. The power of the aristocracy has gone down, as its wealth declined, and industrialization is advancing rapidly—with workers protected by one of the most elaborate social security structures in the world.

The youth are a formidable problem, Doublet thinks—as, indeed, was indicated by the recent convulsion on the streets. There is a greater gap between the generations now, between parent and child, than ever before in Parisian history, in his opinion; of course this is true almost everywhere else in the world. One encouraging factor is the increasing appetite for competitive rough sport. The past generation of youngsters played mild tennis, and seldom went in for group sport; the new generation plays fierce soccer. Taking leave, I asked the Prefect if he thought that the government of Paris was a democracy. He replied, "What an interesting question!" smiled archly, and said no more.

✿

Chief of Police Maurice Grimaud was born in 1913, and received most of his education in Lyons. A subtle, supple, sensitive man, he squints slightly when he faces a hard question, and speaks English with the most delightful of Gallic accents. He has a sloping forehead, hazel eyes, and a good sharp nose, slanting parallel to the forehead. He looked gaunt and a bit drawn when

I saw him. He is imaginative, serious, and conscientious—he even took notes once or twice on things I said about police problems in Tokyo and London, which I had recently been studying. Most observers think that he handled the 1968 riots with signal restraint as well as competence.

Mr. Grimaud saw us in his office on the Île de la Cité, which, seen in photographs from above, looks remarkably like the prow of a ship. He works in a modest building opposite the prefecture; its only elevator is like one in a French apartment house, self-operating and holding no more than two or three persons. On his office walls were a Laurencin, a Chirico, and something in the Picasso manner by an artist I never heard of, Edward Georg. I asked Mr. Grimaud for some figures about something, and he replied cheerfully, "Ah, I have not quite the good memory," and looked the matter up.

Most of Grimaud's career took place abroad until recently; he held various posts in Morocco, Algeria, Germany, and Switzerland. For three stormy years he was Director of Information in Morocco, where he learned to get along well with the press and foreign correspondents. He likes books, and is profoundly literary. Not as much can be said about the chiefs of police in most American cities. He became Prefect of Savoie in 1957, Prefect of the Loire in 1961, and then Director General of the Sûreté Nationale, the national police. Soon came a reorganization merging the Sûreté with the Paris police, and Grimaud has been head of this combined operation for about two years.

The Prefect is, like M. Doublet, appointed to his post directly by the President of the Republic for an unlimited term, and the two men share responsibility for the administration of Paris proper. In addition, Grimaud's jurisdiction includes the three new departments surrounding the city. His budget is $160 million a year, and his staff numbers about 30,000, of whom 19,000 are uniformed Paris cops. Plain-clothes men number between 3,000 and 4,000. "This is a very difficult and complicated city, and we would never let a single person run it," Grimaud smiled candidly. Grimaud has all the routine business of the police, plus traffic and the fire department; public health is divided. Doublet has the hospitals and public services; Grimaud the rat catchers (defense against epidemics), sanitary inspection in the markets and butcher shops, and control of prostitution.

After ten years of service a Paris cop will be earning about $300 per month, good pay by European standards. Many have other jobs after hours—as painters, electricians, plumbers. This is not legal, but is winked at. I asked Mr. Grimaud if the story was true that a few prominent criminals, not yet in jail, sometimes had a close association with high officers of the police, because they were all comrades in the Resistance. The Prefect's answer was a bland "Perhaps." Corruption on minor levels is virtually unknown, because offenders are usually caught quickly and a cop is not apt to risk his job for a small bribe, say for letting off somebody for a parking ticket. The fine for improper parking is petty. French statistics on crime differentiate between the *délit* (burglary, robbery, swindling, et cetera) and actual "crime," like murder or treason. There were 57,313 *délits* in the Paris region in 1966, only 186 actual "crimes"; murders numbered 180.[2] One thing that has worried Grimaud and his men in recent years is, as in London, the sharp increase in robbery with violence. A wave of such robberies began immediately after the Algerian war, when hundreds of thousands of Algerians moved into metropolitan France. There were 1,330 cases of robbery with violence during the first year after Algerian independence; more recently the number has tended to drop off, but it is still considerable.

Race riots? Unknown. Slum conditions? Not good, but slums are few and "specialized." Juvenile delinquency is beginning to become a problem. "There are small aggressions—boys grabbing handbags." A few youngsters steal in order to be able to buy illegal drugs in the black market, but narcotics are not a major problem—at least not yet.

There are three prisons in Paris, two for men, Fresnes and the celebrated Santé, one for women, called La Petite Roquette.

[2] As against more than 600 in New York.

Although the death penalty is still in force, only one execution took place in all France in 1965. The means is still the traditional guillotine, but only three of these unpleasant instruments continue to exist, all of which are housed in Paris. If there has to be an execution in a provincial town, the executioner and his guillotine are shipped there. The executioner's fee is $800, and the present incumbent is, I am informed, a nephew of the previous holder of the post, who "inherited" it from a cousin.

The Remarkable Monsieur Delouvrier

M. Paul Delouvrier, the chief of the planners, is one of the most important men in France, although his name is not widely known. As Prefect of the District of the Region of Paris, he exercises supervision not merely over Paris Intra Muros and Greater Paris but over the whole giant agglutination which includes both of these and constitutes the Paris "District"; this, as mentioned above, has a population of about 8.5 million at present, more than 17 percent of the population of the country. This figure will rise to an estimated 14 million or even more by the year 2000. It is Delouvrier's principal job to deal with this prospect in all its multifarious aspects, and to plan for it. He has to think in terms of thirty years from now. The problem is one of sobering magnitude. You could never mistake M. Delouvrier for anything but a Frenchman. I have seldom seen a more typically Gallic face. Tall, in good trim, with a decisive nose and flashing, deep-brown eyes, he seems younger than his age, fifty-three. He has a modest, direct manner, and speaks with lucidity and grace; his suavity conceals power. He has a distinct quality of dash as well. De Gaulle said of him once appreciatively, "Ah—a man who believes in solutions!"

A man from the mountains, Delouvrier was born in the Vosges and has been called the "perfect technocrat." He told us, smiling, "Ah, but I am not that, because I do not speak English!" He became at an early age an inspector of finance in the civil service —a job which exists in no other country, but in France goes to none except the most skilled and brilliant candidates for the civil service after a wrenchingly difficult examination. The war came, and after brief service in the army Delouvrier became a Resistance fighter leading a detachment of *maquis* in the Fontainebleau area. Resuming civil life, he rose rapidly, and reached the important post of director of the cabinet in the Ministry of Finance in his early thirties. Then he held other similar posts, worked for a time with the European Coal and Steel Community (part of the Common Market now), became addicted to the planning concept, taught city planning and urban renewal at the Institute of Political Studies in Paris, and served briefly as Governor of Algeria, where his record was mixed. Politics kept getting in his way. When he relinquished the Algiers office, de Gaulle offered him the job of "reconstructing" Paris.

"This is a good post. What will you need?" de Gaulle said.

"I don't know," responded Delouvrier.

"I don't admit that you don't know," de Gaulle replied.

Delouvrier bought a secondhand car and set out to explore Paris inch by inch. What he saw appalled him—congestion, apathy, an anarchic lack of cohesion. Similarly, he had been much struck by urban difficulties in Algiers. He tried to learn by Haussmann's example, and started to read the great man's autobiography—he gave up after seven pages. This was a new world now, and the central problem was baffling in a new dimension— how could the enormous spreading mass of Paris be organized into an effectively functioning metropolis? *It is impossible to keep a modern city from growing, but how control its growth?*

Delouvrier picked up a suitcase and, with two colleagues, went around the world inspecting its chief cities: New York, Philadelphia, Detroit, Moscow, London, Stockholm, Tokyo, Rio de Janeiro, Helsinki. The Scandinavian cities were those run best, he thought. But Paris was bigger, and had to be dealt with from a different point of view. Here lived several million *French*-men —"anarchists." Delouvrier got to work. This was in 1961, and he has been the major personality associated with Paris planning and development ever since.

The Prefect has modest offices in the Rue Barbet-de-Jouy, a charming small street on the Left Bank near the Rue de Grenelle and the Rue de Varenne; this was once the most fashionable neighborhood in France intellectually and otherwise, corresponding to Bloomsbury in London. Close by are various ministries and embassies, most of them converted from private mansions —*hôtels particuliers*—in the eighteenth century. The official residence of the Prime Minister, the Hôtel Matignon, is around the corner, and so is the Archbishopric of Paris. Much of the aristocratic, wealthy element here has tended recently to move across the river. The basic flow of Paris today is from Left Bank to Right, but this crystalline Left Bank enclave remains unspoiled and virtually intact.

After brief amenities Delouvrier, a busy man, led us across his office and, briefing us, demonstrated a series of large maps and charts. His work has many aspects—highways, parking areas, air pollution, regional express railways, water supply, housing, schools—but what he chose to emphasize was the central concept, *aménagement du territoire,* which may be translated informally as "environmental planning." Every aspect of the work is part of a scheme planned down to the most minute detail. But of course planning in relation to Paris began long before Delouvrier. Haussmann was a planner. The idea of *regional* development applied to the Paris area first became seriously considered in 1957-59, and the District of the Region of Paris was created, against some lively opposition, in August, 1961, to achieve this goal. Soon Delouvrier had ready a document estimating the region's population and industrial growth for the next generation, and actual projects were launched the next year under a four-year emergency program. Then in 1966 a Ten-Year Plan was set up with more specific goals, the cost of which will, it is estimated, exceed $60 billion over the next thirty years.

Delouvrier does not want to build satellite towns, like the "new towns" surrounding London, which he thinks hinder development more than they encourage it, and which he calls "sterile" and "joyless." The Paris concept is based on growth, but not at the cost of human values, at least in theory. Nor does he favor the "spontaneous" creation of new communities. "We do not want to grow like inkspots spreading." The elaborate and ambitious mechanism chosen for Paris is the construction of two new "tangential axes," which will run all the way to the English Channel parallel to the Seine. One will have its exit at Le Havre, one at Caen. The great parks, Fontainebleau, Rambouillet, and so on, will automatically give them a background of green near Paris, more satisfactory than the Green Belt in London; the land is owned by the state, and is inalienable. One of these new gigantic axes will swoop around Paris in a wide turn and follow the Marne to the south. Each will have its own express railway and superhighway system—"veins" relieving congestion and giving mobility and lifeblood to the new metropolis.

Once these axes are in place—perhaps the planners are too hopeful—eight new cities, called "hubs," will be built at appropriate intervals along their routes a few miles from the Seine, with populations ranging from 300,000 to a million—a vast conception. The whole of the Seine basin, stretching all the way to the Channel, will thus become an integrated extension of the Paris area. Paris will resemble a giant sleeve. The new "hubs" will be real cities, not outposts, "with a complete range of urban components." Delouvrier added dryly, "Of course we cannot give them all a Comédie Française, but we will try to give them practically everything else—good housing, industrial plants, every cultural advantage." These cities have already been located and named. Although separated geographically, they will all be considered to be *part of Paris,* so that nobody need feel exiled from the capital. The main desideratum is not the creation of new communities but the extension of Paris itself so that it will be able to master its inevitable growth and be able to house comfortably its anticipated population of fourteen million by the year 2000.

Certainly the need is pressing—even now. Perhaps Delouvrier, Grimaud, Doublet, and company think too obsessively in terms of the future, but they do not neglect the realities of today.

There is certainly a great deal to do in the present. I know few statistics more appalling than some having to do with contemporary Paris. Sixty-two percent of the population has no bathtub or shower as of recent date, 60 percent no central heating, 15 percent no electricity, 49 percent no indoor toilets.[3] More than half of all Paris housing was built before 1915, and is clearly obsolescent. The Delouvrier plan calls for no fewer than 145,000 new housing units a year within the Paris region—a total of 1.5 million by 1975. Not only are the eight new "hubs" to be built along the Seine, but six great urban centers within metropolitan Paris itself are to be renovated—including Défense-Nanterre, St.-Denis, Bobigny, Créteil, and Versailles. We spent a day watching the work at Nanterre—a new city is being erected here, inch by inch. One telling detail is that Les Halles, the central market which has been one of the traditional monuments of Paris for generations, is to be moved bodily to new quarters on the outskirts of the city, to ameliorate congestion and give room for new fructifying development.

[3] Henry Giniger in the *New York Times,* February 26, 1963.

4. MORE ABOUT
LA VIE PARISIENNE

Superior in rank to any of the triumvirs I have just mentioned is André Malraux, the novelist and art historian who has been close to de Gaulle for many years and who is Minister for Cultural Affairs. As a member of the cabinet he is a national rather than a merely Parisian figure, but most of his work pivots on Paris and he has contributed greatly to its renewal—not merely as a planner, but as a kind of cultural director at large. He is the man chiefly responsible for cleaning up much of Paris and for its spectacular illumination.

Malraux is difficult to see—almost as difficult as de Gaulle. He is a busy as well as complicated man, hard at work on several new books, following his autobiography, and prefers in any case to be inaccessible.

But an interview was arranged for my wife and me promptly through the intermediation of an old friend at the Quai d'Orsay. As a matter of fact, I have met Malraux several times in cordial circumstances, but I was not sure that he would remember me. Back in the thirties I superintended a public dinner in New York for Ernest Hemingway when he, Hemingway, was about to take off for Spain to cover the Civil War, and Malraux was a principal guest. He had just returned from the Spanish front, where he, too, had been a fighter for the Loyalist cause.

Now, a generation later, Malraux has gained weight and has shed much of his former nervousness. He looked sleek. He received us in his quarters in the Palais Royale, the traditional residence in older days of the Duc d'Orléans, the King's brother. The terrace overlooks a broad court, which happened at the moment to be housing an open-air exhibition of modern sculpture, and where stout, impassive Parisian women, with the marks of years of effort and fatigue engraved on their faces, sat on benches knitting in the pale sun. They probably did not have a sou to spare among them, but at the far end of the court is one of the most luxurious and expensive restaurants in the world, the Grand Véfour.

Malraux began by reminiscing about the Hemingway dinner years before. It marked the first time Hemingway had ever worn a dinner jacket, which, Malraux said, he rented for the occasion. Malraux took us out on the terrace and pointed to a window which marked the residence a century ago of Alexandre Dumas. He was pleased by the elegant aristocracy of his own quarters, where Louis Philippe Napoleon III, and the Empress Eugénie had once lived, and took us back inside. The decor is an intriguing mixture of two periods—gilt carving of the eighteenth century on the doorways and arches, and walls painted with flowers installed by Eugénie a century later. Jérôme Bonaparte died in this room, which at that time contained little but mirrors and an enormous bed. Malraux showed us, among several glistening artifacts, an example of Cretan sculpture, the marble figure of a cat, paws up ("to catch luck"), and a lamp—also a candlestick—that had belonged to Napoleon I, marked with an imposing "N."

He proceeded to satisfy the object of my visit, telling us about his work cleaning up Paris, washing its face. The classic monuments of the city glow these days with a new revealing light, and the Arc de Triomphe, as an example, is no longer a dirty gray but the color of Devonshire cream. Malraux worked out this project with a contractor whom he met by accident. The official architects had all said that the job of cleaning the ancient monuments was impossible. Then a man unknown to him happened to send him a complete set of his first editions, asking that they be autographed. Malraux signed them, and the man thanked him with a letter saying that he, Malraux, must be a "very good fellow" to have performed this chore "without asking any question." Malraux's curiosity was piqued, and he found out that the man was a contractor. They met, and Malraux entrusted him with the clean-up job. There were subtleties and intricacies aplenty. First, a method had to be invented by which the crumbling old stone could be swabbed without damage. Sponges were used at first, then brushes under a continuous flow of water without detergents. A few buildings which could do well with a cleaning, like Notre Dame and the Cluny Museum, have not been touched because their scrolls of stone are too fragile; they might crumble into dust, but an attempt will be made to deal with Notre Dame soon. Malraux was very eloquent about all this. What the clean-up did was not merely reveal the original whiteness, but also the dark—the shadows. "What we did," he told us, "was to show the dark."

I might add a word about other things Malraux talked about. He discussed the position of some younger French writers and artists and did not appear to have a high opinion of them. He said that they were not interesting politically, because they did not cope with the basic problem of modern society, which is how to achieve a workable, viable system now that Marxism has become obsolete. The novel is dead. He himself had written none in twenty years and Hemingway, as another example, wrote only two short novellas in the last two decades of his life. Veteran French writers who should be at the height of their powers, like Aragon and Mauriac, are played out. Similarly, no new generation of artists is coming up. "Young" painters, like Dubuffet, are in their sixties. I asked him the question to which I have given several answers in the preceding chapter—"What accounts for today's intellectual apathy?" First, he thinks, an appalling inner sterility which began to express itself in 1937 or thereabouts. Second, the breakdown of the bourgeoisie. Before the war women stayed at home, with nothing to do but read; in consequence writers had wide audiences, they were given a serious reception. Today in an era of two-earner families women don't have time or opportunity to read or appreciate art, and stimulation to the writer or artist is cut off.

Malraux's interests and responsibilities are, as befits his character, voluminous. He was the originator of the idea of restoring the Grand Trianon Palace at Versailles both as a museum and as a guest house for visiting heads of government; he has built *maisons de la culture* (cultural centers) in half a dozen cities all over the country in order to decentralize cultural activity; it was he who commissioned Marc Chagall to do the new ceiling for the Paris Opéra. I mentioned that Chagall had also contributed to Lincoln Center in New York. "Ah, but you always copy from us," Malraux replied blandly. Then he arranged to lend us his box at the Opéra one evening so that we might see this radiant work of art for ourselves. (The performance was an effective but mortally long *Don Carlos*.) We admired the Chagall ceiling, but I thought irreverently that it looked like a cross between Bemelmans and Dufy, with a green moon, a large red apple (or was it an apple?) pierced by a sideways-leaning Eiffel Tower, a yawning seashell, yellow angels, and dolls riding cocks.

Briefly, as our talk continued, Malraux touched on politics and other items. He scorned the eggheads. "If you really want to know about political thought in this country, talk to your postman." (But, in a sense, this has always been true of Paris.) Nobody is interested in theory any more; the things that truly concern citizens are how to find a new apartment, how to fight their way through traffic jams. "Paris was no more built for the automobile than was the district around the Battery in New York." There are no political parties in a real sense any more. France and Paris, he reminded us, were totally different things, even though Paris "is" still France. The countrymen despise Paris. When a car comes along with a license plate indicating Paris, they mutter, "Dirty Parisians!"

Then he told us not to miss visiting the Marais, where we would see the Paris illumination at its best. This district in the Second Arrondissement, once fashionable, has degenerated into a slum, but half a dozen of the marvelous old palaces and mansions survive intact. Literally, "Marais" means "swamp." We drove through the area at night gazing raptly at the façades of the Soubise, the Sully, and the Carnavalet palaces, among others, all shining with a silver luminous glow. Another night we went to see *La Surprise de l'Amour,* an old Marivaux play, in the courtyard of the Hôtel de Rohan, the showpiece of all the palaces, now lit by a similar creamy phosphorescence. Slim Doric columns give boundaries to a large semicircular stage, and the audience, solidly banked in an amphitheater, seemed big enough to fill the Yale Bowl. A soft summer wind was blowing, and leaves flapped through the blurred golden haze above the stage; I could not tell whether they were real or props from trees set up to give background to the play.

✿

After the great public monuments, like the Madeleine and the Chambre des Députés across the Place de la Concorde, were scrubbed clean, a ruling went into effect making the cleansing of private homes obligatory in several neighborhoods, to be paid for by the landlords. I met one who said that to clean his property had cost him nearly $30,000, but it had become white as a pearl, and he was proud of its restored beauty. A novelty in this field is that anybody in Paris can have any building especially illuminated at any time—say, as a gesture to a friend or to celebrate a party—for a modest fee.

Paris is, incidentally, the only major capital in the world the streets of which are cleaned by hand daily. Every morning the hydrants are turned on and water flows along the gutters while an army of 4,200 street cleaners sweep the pavements. These workers are mostly from Algeria, Tunisia, Morocco, Senegal, and other former French colonies; their wages are the equivalent of about $100 a month. Then, too, there are over 8,160 garbage men in Paris who pick up more than a million tons of garbage a day. Metal garbage cans are forbidden, as in London; only rubber or plastic cans may be used, in order to diminish noise.

Sights, Sounds, Symbols

Paris has two types of slum, it seems. On the outskirts of the city are several *bidonvilles*, or shantytowns, which are more or less segregated by nationality, like the Algerian community near Nanterre, where some Moslem women still wear veils. Another is located all over the city on the *upper* floors of old mansions, where Spanish and Portuguese families mainly live, inhabiting former servants' quarters—often without light, heat, or an adequate water supply. The roofs of the city form, in fact, a continuous slum, a kind of ceiling, over wide areas. The extent of the influx of foreign immigrants is not always appreciated. One little joke is to the effect that something astonishing happened at the new Hotel Hilton the other day—a servant was encountered who actually spoke French.

At the other end of the gamut high society in France, particularly Paris, is still probably the most closed, arrogant, and impermeable in the world. At an official dinner—or even at a purely social affair—the strictest rules of protocol are observed. Here is a city where rank does count. The Paris of the grand old families still lives in the grandest style. First priority at a formal dinner goes to a member of the Académie Française, who must be seated higher than anybody else—an interesting survival of French respect for the intellect; after that come ambassadors and holders of the Grand Cross of the Legion of Honor. (Incidentally, it is considered somewhat chic for a member of the Legion of Honor *not* to wear his red ribbon or rosette in his lapel in informal or business dress.) Incorrect seating at a dinner party may cause humiliating difficulties for an erring hostess. A guest improperly placed may go so far as to show his displeasure either by leaving or by turning his plate upside down and refusing to eat. This has happened even in French embassies abroad.

Men and women are never separated and herded into different rooms after dinner for a brief period as is the custom in social circles in London and New York. All go into the drawing room together, but men and women do tend to cluster into different groups. No one in society, male or female, *ever* asks to go to the powder room. This is considered to be the height of indelicacy. Another odd point is that persons are addressed by name only if they are fairly intimate. The rule is to say simply "Monsieur" rather than an actual name. An old maxim has it, "If you call a man by his name he'll think you're a tradesman." The universal and uninhibited use of Christian names, particularly on first acquaintance, as in America, is considered horrifying. Parisian society can, in a word, be extremely stiff. We went to one party where cabinet ministers talked chin-to-chin. My wife sat next to a man of distinguished lineage who denied the validity of any experience, almost like a character in Proust, and who kept on saying in a bored manner, "But what does any of it *matter?*" This is why France fell.

The foremost preoccupation of the middle class, a large majority of the population, is money. We had drinks one day with a young man whom we have known since his teens, and who is now earning a living as the second concierge in a leading hotel. "Paris is exciting only if you have plenty of money," he averred somewhat glumly, adding that it was difficult for him to get a decent meal for less than nine francs ($1.80), and that rent in a miserable building took almost a third of his income. Yet, expensive as it is, one of the principal satisfactions of Paris is that, if necessary, a resident or tourist can live here at almost any level, even the lowest, without undue suffering. Being poor in London or Berlin is much less pleasant than in Paris. No matter what, Paris has its *joie de vivre.*

One evening we took another young man, the cultivated seventeen-year-old son of a French ambassador abroad, to supper in a café; we talked not about money but about education, and he sought patiently to explain to us some characteristics of Parisian schools. One outstanding quality is that children from the earliest age are trained to be responsible. Most students are intensely serious, work hard, and seek above all to save for their holidays. They are younger than students of the same class were five years ago. Instruction at the average *lycée* is rigorous, and to fail or drop out is a disgrace. "Failure can terminate a life." French universities are, however, lax. Nobody pays attention to whether a student attends classes or not, and, as mentioned in the preceding chapter, methods of pedagogy are stultifyingly archaic. The Sorbonne, founded in 1253, is the fourth oldest university in the world, after Bologna (1050), Salamanca (1218), and Padua (1222). With nearly 100,000 students today, it has been forced to decentralize, which was one reason for the 1968 uprising. The Faculty of Letters moved out to Nanterre, the Faculty of Science to Orsay, the business school (Hautes Études Commerciales) to Jouy-en-Josas, and the Polytechnic Institute to Palaiseau. This dispersion has, incidentally, irremediably changed the atmosphere of the Latin Quarter, the original site of the Sorbonne. This is no longer predominantly a district dominated by students, artists, and Bohemians, but is becoming a kind of second-class bourgeois shopping center.

Some average incomes in Paris are:

Unskilled worker	$2,185 per annum plus overtime
Skilled worker	$2,512 per annum plus overtime
Teacher	$4,250 per annum
Postman	$1,380 per annum
Bank employee	$2,600 per annum
Administrative framework	
a. Middle echelon	$4,605 per annum
b. Upper echelon	$9,200 per annum
Army major	$5,200 per annum plus representation and housing allowance
Deputy	$14,500 per annum

And prices:

Kilo[4] of potatoes	5 to 15 cents depending on season
Kilo of butter	$1.90
Kilo of sugar	28 cents

[4] Approximately 2.2 pounds.

Liter of milk	16 cents
Kilo of best-grade beef	$4.00
Woman's coat (of decent quality)	$80.00 up
Man's suit	$80.00 up
Children's shoes	$6.00 to $15.00
Rent, one-room apartment	$120.00 per month (unfurnished, not under rent control)
Transistor radio	$20.00
TV set	$287.00
170-liter refrigerator	$1,520.00
Peugeot 404	$2,190.00

Who owns Paris. Until the great French Revolution of 1789, most of the city was a cluster of homes, tenements, shops, and individual dwellings for rich and poor, all privately owned, but this is not the case today. Paris has about 70,000 buildings at present, of which the municipality itself owns about 7,000, or 10 percent; some 800 belong to the French state.

After these, the largest real-estate owners are not private families or individuals, as in London, but large companies (many of them nationalized), banks, and insurance companies. The Société Gérance de la Ville de Paris, which is half-owned by the municipality, has very large holdings; one of its subsidiaries, known as SAGI (Société Auxiliare de Gestion Immobilière) owns 20,000 lodgings in 150 to 200 apartment buildings. One big insurance company, the Séquanaise, owns about 150 apartments; the Union National sur La Voie has 100, most of them on or near the Boulevards Militaires on the edge of the city.

The leading banks, like the Crédit Lyonnais, usually own the buildings of which their branches occupy the first floor; the Crédit Lyonnais has 81 such branches throughout Paris, the Société Générale 66. The biggest private owners are the Rothschilds and several aristocratic families like those of the Baron Louis La Caze (20 apartment buildings), the Duc and Comtes de Montesquiou (12), and the Marquis and Comtes de Buisgelin (14). The Rothschilds own 36 different properties.

What runs Paris? I asked this question of one local journalist, who replied, "The green line [hot telephone] between de Gaulle and the Kremlin." I said that I thought this was nonsense, and still do. Paris is run by de Gaulle, the survivors of the Resistance, the bureaucracy, the clergy to an extent, the propertied class— and rebellious students if they get a chance.

✻

Paris is changing to a limited degree in several fields, but much remains the same. There seem to be fewer *vespasiennes* (public urinals), and butcher shops marked with a horse's head (to show that horse meat is available) are not, I think, as numerous as before. An outcry about these has occurred because British and Irish horses are shipped here by boat to supply the French market, and the seas are often rough; a horse is prone to seasickness but cannot vomit, and therefore, according to British humanitarians, suffers horribly before it is slaughtered.

Waiters in the cafés do not, if my superficial impression is correct, total the *addition* by taking note of the figures on the saucers accompanying each drink, as in former days. There are, incidentally, 13,977 cafés, or "terraces" as they are properly called, in the city. The mail service by *pneumatique,* whereby a letter will reach any destination in Paris within an hour or two, still functions admirably. Regular mail is delivered three times a day, more than can be said of New York. An emergency telephone service gives advice on medical care in the case of accidents, and a "Suicides Anonymous" exists, as in London. Even so, it is a remarkable fact that there have been 350 suicides in Paris by persons jumping off the Eiffel Tower since it was built in 1889.

Miniature-auto racing is a contemporary fad, and so is bowling. Sixty-nine families (237 people) live in the Bois de Boulogne, the heads of which are policemen or civil servants housed in pavilions belonging to the state, and where police dogs are trained. The chic prostitutes in the Vendôme area, many of whom wear tight white slacks and brightly colored woolen shirts, drive around in sports cars, and charge 500 francs ($100). Wives in suburban homes often give themselves to prostitution in the afternoons, but, on the other hand, the traditional *cinq-à-sept* (lovers' trysting time) has, I heard, more or less gone out in Paris itself because traffic jams make it difficult to get around between five and seven.

A trend exists for bourgeois families to live in "ranches" just outside the suburbs; men of the household even go in for cowboy hats. One evening—if I may continue to list a few curiosities —I dined with a lady who had on view two portraits of herself by Picasso, neither of which showed her face; this was not because she was not attractive, but because the artist thought that her hair was so pretty that he would limit himself to a back view.

I did not know before that Paris newspapers are delivered by mail, not by truck or newsboys, and that this is the reason why Sunday newspapers (except one) do not exist, since no mail service exists on Sundays. Nor had I ever heard much about the social security system, which is rigorously administered. An average employer is obliged to pay about 45 percent of his employees' wages into the security system.

Paris, a green city, takes particularly good care of its trees, of which there are more—some 85,000—growing out of the pavement than in any city in the world. A private memorandum tells me that they line 300 kilometers (186 miles) of streets, roughly the distance from Paris to Brussels, exclusive of those in the parks, which, incidentally, are extremely well kept up (as are the three great cemeteries, the Père-Lachaise, Montparnasse, and Montmartre). One celebrated street, the Avenue de l'Opéra, has no trees, because the architect who built the Opéra building itself insisted that nothing should obstruct the view. Trees planted in Paris have to be straight, so as not to take up too much space, with branches going up, not down; they are trimmed in a particular esoteric way like one-sided umbrellas, and they may not bear fruit for fear of injuring passers-by. True, horse chestnuts, which do insist on shedding chestnuts, are still numerous. The plane tree is the tree most common all over Paris, but there are elms, poplars, maples, sycamores, and catalpas as well. The trees are so chosen that most streets will be green from early spring straight through to November.

The chief problem of the city is, on a day-to-day level, traffic, but the congestion is not quite so murderous as local citizens say; it is much worse in Rome. Still, it took me forty-two minutes on one occasion to drive by taxi from a point near the Étoile to the Hotel Ritz, a distance of a mile or so. There are 1.8 million cars in Paris and the suburbs, of which the city is capable of handling 300,000.[5] Taxis are not, incidentally, so scarce as is generally thought except, as in New York, at certain special hours in certain areas. Many taxi drivers are women, and some are novices. I rode with a fierce virago one afternoon who did not know where the Quai d'Orsay was, and who, in a stern voice, commanded me not to smoke in her taxi, because she was "*allergique au tabac.*" A taxi picked up near a railway station is—not many visitors know this—entitled to a franc extra on the fare.

In Paris I put up at the Ritz if I can, and nothing can be more satisfying than this superlative hotel. It is stamped from top to bottom with its own regal, discreet, fastidious, and sumptuous character. What other hotel has ceilings that seem to be twenty feet high, and is so improbably organized as to have two different second and third floors with room No. 60 directly opposite No. 131? Where else is there a fourth call button at your bedside, colored blue, in addition to those red, yellow, and green for waiter, maid, and valet, to summon your private servant (if you have one), and where else do you turn the lights on with a key and tell the time from a noiseless golden clock set in a brocaded wall? The long corridor of *vitrines* between the Vendôme and Cambon sides is, I think, the most glamorous shopping "street" in Europe. The elevators open on two sides, like those in the Gritti in Venice. Some light switches are double. The telephone

[5] Lloyd Garrison in the *New York Times*, December 23, 1967.

girls know the number of the American Embassy by heart, but not of the British. The beds, bed linen, and toilet accouterments are of the first quality. I found that George, the chief barman on the Cambon side, a historic character, and Robert the concierge, equally so, had retired since my last visit three years before, but they have worthy successors. One morning I insisted on buying a stamp and stamping a letter myself instead of leaving this to a junior concierge. He rebuked me with a polite "Ah, no confidence!" A black mark against me! (In some European hotels—never the Ritz!—letters posted via the concierge have been known never to reach their destination, because the bellboy throws away the letter, en route to the post office, in order to steal the stamps.)

Our day usually began when we left the hotel on the Vendôme side in midmorning and watched the Cadillacs trying to twist their long bodies into position on this noble square. The Place Vendôme has been nicely slicked up in recent days, and I had never been able to see before that the balconies are painted a dark royal blue, not black. I did not know until a friend told me that a heroic statue of Louis XIV once stood in the place now occupied by Napoleon's column here, and that the column itself, more than 120 feet high, is made of bronze cast from the cannon captured at Austerlitz.

A waiter, immaculate in a white coat, scooted across the Rue de Castiglione, bearing a large silver tray on which reposed dishes and tureens. Presumably he was transporting an early lunch from the Ritz to a customer stranded in a shop. I thought that this was somewhat odd, but nobody paid the slightest attention. You can do almost anything in Paris, and no one will pay attention. A day or two later we ran into a wonderful old drunk near the Louvre, too intoxicated to stand, who seemed about to *roll* himself across a street boiling with traffic. The corner cop did not raise an eyelash.

But some untoward incidents do have to be dealt with. We had lunch with Ambassador Bohlen at the Embassy residence; he did not mention it, but a madman had attempted to assault him with a knife as he entered his car the morning before. The Paris chief of police, the estimable M. Grimaud, telephoned Bohlen to explain and apologize after the assailant had been arrested.

"The man is crazy. He has only one *idée fixe*. It outrides all."

"What is it?" Bohlen asked.

"To kill you."

I sauntered down the Rue du Faubourg St. Honoré, with its suave, immaculate shops, and glanced at No. 55, the Élysée Palace where de Gaulle lives and works. All sixteen Presidents of France since Marshal MacMahon in 1874 have lived here. The building has a romantic history. It was built, I heard, some 250 years ago by Henri-Louis de La Tour d'Auvergne, Count of Évreux, and bought on his death by Madame de Pompadour, the mistress of King Louis XV. After decorating it with ornate splendor, she bequeathed it to the monarch. Presently he sold it to a businessman, who, a few years later, sold it back to the crown at a large profit. The poet Alfred de Vigny, Prince Joachim Murat (who married Napoleon's sister Caroline), and Napoleon himself lived here at one time or other, and it was the scene of Napoleon's abdication after Waterloo. The palace has a Salle des Fêtes, nicknamed "the Station," which can hold ten thousand guests at a ball. One room where Prince Louis Napoleon engineered the coup in 1852 which made him Napoleon III, called affectionately the Salon d'Argent, is kept inviolate and is never used for any functions. For a time the palace was defaced by an immense glassed-in hall, known as the Monkey's Palace, but this was removed by President Vincent Auriol in the 1940's. An entrance to the gardens on the Champs Élysées side, the Cock's Gate, is utilized only for welcoming heads of state. De Gaulle has vastly changed the character of the Élysée by making it a genuine working office, like the White House. Several hundred officials perform their duties here every day. A document tells me that the personal staff is large, including twelve presidential secretaries, Madame de Gaulle's private secretary and maid, two doctors, a nurse, the chief housekeeper, thirty chauffeurs (for fifteen cars), seven cooks, six valets, four butlers, ten ushers, and such maintenance men as the four silversmiths whose sole duty it is to keep the silver polished, five laundrywomen, and two *lustriers* who preserve the condition of the chandeliers, of which there are one hundred.

✿

Paris is of course the world's capital of the fashion business, although in recent years Rome has been pressing it. Altogether there are about six hundred couturier houses here. The most celebrated and successful at the moment are Givenchy, Balenciaga, Dior, Marc Bohan, Balmain, Yves St. Laurent (a Dior offshoot), and Courrèges. Historic figures are Vionnet, who invented the bias cut; Poiret, who "denationalized" French fashion and was much influenced by the Russian ballet; Chanel, who abolished corsets and the like, thus giving women a better chance to breathe; *enfants terribles* like Schiaparelli, and above all Dior, who created the New Look. Dior signified romanticism, although his work was solid as a battleship. He believed in petticoats and hooks and eyes. He let skirts down. One reason was the aftermath of the war when—by 1947-49—enough textiles had become available to make skirts of any length, after a prolonged period when they had to be short because so little material was to hand.

What makes fashions change? Why should shoes be pointed one year, blunt the next? Who exactly decides whether there should or should not be a waistline in a dress? This is a mystery. The impetus may come from an adventurous designer or even from an important individual (and necessarily wealthy) customer among the fashionable who demands something "different," and is in a position to reward a couturier who can provide it.

Why is Paris the world's fashion center? One reason is that it is a city which has always had a civilized wealthy class giving encouragement to beauty in craftsmanship (consider Limoges and Sèvres); another is that it is full of expert creative craftsmen in small accouterments—weavers, button makers, glovemakers, leather specialists, and other manufacturers of accessories. Then, too, Paris itself gives an "inspirational" background of balance, proportion, taste, restraint. But nowadays the young people are breaking down some inhibitions, because they want "freer" clothes, less style, more comfort. Paris has its Mods, just like London. A taxi took me down a Left Bank street one afternoon and the driver saw, as I did, a marvelously built young Negro woman with the proud, graceful movement of an African queen, wearing the shortest miniskirt I ever looked at. I admired this sight, but the driver grumbled, "Paris has become a brothel!"

One day we went to call on Man Ray, the veteran American painter and photographer, who, at seventy-seven, is still as inventive and copious as he was forty years ago. At least six different art worlds exist in Paris, he told us, from the "rich French" (meaning Dubuffet), to the youngsters "who like to swing bats." Man Ray showed us with amusement some *square* dumbbells he had just designed.

Then we sought to investigate contemporary aspects of the literary world. Friends told us that several leading writers today call themselves "structuralists"—a seemingly inappropriate term in that the current mode in fiction is to be formless, negative, with no strong story line, and with the chief emphasis on the way a thing is said, not on what it says. Of course this is nothing new. One intellectual influence of consequence is that of Jacques Lacan, a sixty-six-year-old linguistic scholar and psychiatrist; other important figures are a fairly youthful essayist, Jean-François Revel, the well-known anthropologist Claude Lévi-Strauss, Alain Robbe-Grillet, Michel Butor, and Nathalie Sarraute.

Some Satisfactions in Gastronomy

What is most fun in Paris is, of course, going to restaurants. French families practically never have a foreign guest to their homes; the home is an inviolable hidden cave. Restaurants are, as is universally known, taken with extreme seriousness, and their standards are probably the highest in the world. The plain business of mere eating is a serious matter here. I knew a young American girl who, living in Paris, had a French lover; it shocked her for some little time that, during meals with her, he never spoke a word. All of his attention went to the food. I had one lunch with a Parisian colleague that began at 1:15 and ended at 5:45. It was sublime. It was not worth it. My wife and I like to have one meal at the Grand Véfour as soon as we arrive in Paris, then one at Laperouse. After this we choose other restaurants alternately. I like bistros better than the grand establishments. One of my favorites is Chez Pauline, near the Opéra; it has scarcely a dozen tables, but the cuisine is of the soundest. Similarly, I would hate to be in Paris and miss a meal at Raffatin et Honorine on the St.-Michel end of the Boulevard St.-Germain. Its proprietors invented the idea that a large miscellany of desserts in the manner of hors d'oeuvres might be offered at the *end* of a meal so that the customer could nibble at a dozen or more different sweet things instead of merely one. The idea caught on. If my memory is correct, this admirable small establishment once had a scale on its terrace; you weighed yourself going in and out, and if you had not gained a pound or two, the amiable proprietors were aggrieved.

The two "in" places now aside from the standard classic restaurants appear to be Chez Garin and L'Orangerie. Both have the atmosphere of a bistro, but Garin is fancier. In both, guests are greeted by a young woman, presumably the daughter of the owner or chef, instead of a maître d'hôtel. A specialty at the former is a thin steak rolled in a crêpe; at the latter, where there is no menu, you are served a huge basket of *crudités* and then some such fantasy as *écrevisses* with scrambled eggs, followed by a variety of roast meats on a plank. The most beautiful dining room I have ever seen is the shimmering silver main restaurant of the Hôtel Plaza Athénée. The whitest thing I know in the realm of food is turbot at the Ritz. A noteworthy processed cheese is Boursault, to be found at Chez Allard, a well-known Left Bank bistro, and elsewhere. At the Grand Véfour, the freshest of sweet peas decorate the tables, and soft marzipan cookies of incredible subtlety of flavor come with the dessert. An *apéritif* well worth trying if you are tired of hard drinks is a *kir*, made of cassis and white wine. I was told that it was named for a mayor of Dijon.

An ancillary sport is chasing stars in Michelin. This guide, as incorruptible as *Variety* in New York, gives annual ratings to about 300 of the 10,577 restaurants in Paris. If a restaurant deteriorates for some reason and loses a star, this can be a very serious matter; one of the owners of the Relais de Porquerolles, a popular establishment, killed himself last year when Michelin withdrew its two stars. Michelin gives its supreme accolade—three stars—to only twelve restaurants in the whole of France; five of these are in Paris. There are sixty-seven two-stars in the country at large, eighteen of which are in Paris. The five three-stars in Paris are Maxim's, Laperouse, La Tour d'Argent, Le Grand Véfour, and Lasserre. I asked a good friend who owns one of these if he knew it when a Michelin inspector dropped in for a meal. Never. The anonymity of the Michelin men (there are believed to be ten traveling inspectors in all) is sacrosanct, and they never expose themselves unless they want to ask questions *after* a meal is served.

The most fashionable discotheques at the moment are Castel's and the New Jimmy's. All over Europe these flourishing establishments are as similar as white mice. Harry's New York Bar on the Rue Daunou, once the exclusive haunt of American newspapermen and other barflies and visitors from the States, where the French language was scarcely even heard of, has become practically monopolized at night by Parisian café society, strange as this may be. Copies of British pubs have become popular, like the Pub Churchill near the Étoile and the Duke of Bedford's Arms on the Left Bank. The Crazy Horse Saloon on the Avenue George V, is still, I think, the best strip-tease joint in the world. The accent at present is on bottoms rather than breasts, and one dancer bears the nice name Bettina Uranium.

. . . *And to Conclude*

We repeat the question: Is Paris dead? The answer is clearly no, although the city shows a good deal of apathy, disillusion, and lack of spark, the 1968 riots to the contrary notwithstanding. Two major factors help to give a more optimistic and challenging note, and, strangely enough, they interlock. One is the planning concept which is being applied to Paris as assiduously and with as expert skill as in any city in the world. The other is the city's incomparable, supernal beauty, together with its quality of mysterious, attainable romance. Paris can still generate emotion in the visitor almost as a beloved person can; to many it is almost the equivalent of an actual person—brother, mistress, friend. In some respects, this City of Light is still the center of the world, more thrilling and magnetic than any other capital, proud, chic, and occupied by several million coldly passionate individualists who still, even in the humblest station, express the unconquerable *gloire* of France.

5. THE CIVILIZATION
OF FRIED POTATOES

Belgium has been a carpet for invaders and marauders since the
beginning of time, but Brussels, its sturdy bourgeois capital, has
become recently a symbol of the development of European cohesion and unity. With all its contradictions and peculiarities, which
are marked, Brussels rises out of a past beset by conquerors to
become the nearest thing we have to an international capital of
Europe. Both the Common Market and NATO (the North Atlantic Treaty Organization) have their headquarters here. Yet,
although it represents several major elements in the struggle for
European integration and federation, Brussels is a city harshly
riven within itself.

People call Brussels dull, and in a way it is—sober, stolid,
middle-class, with little charm. When a Frenchman here was asked
what Brussels was like compared to Paris, he replied, "It's like
going out with the sister of the woman you love."

Then, too, people say that the character of Brussels is to have
no character. The truth is that it has plenty of character, but of a
kind hard to define because it partakes of several contrasting
elements, some of which seem to cancel each other out. Partly
this is because Belgium is a bilingual country deriving from
both Teutonic and Gallic strains.

Brussels is phlegmatic to a degree, materialistic, standpat, and
industrious. But it is also capable of the most astounding insta-

bilities and irrationalities, and it has fiercely exacting problems to which citizens respond fiercely. Scratch even the most sober-sided Belgian, and you will get the blood of a *frondeur*.

Voltaire wrote of Brussels once:

> In this sad town wherein I stay
> Ignorance, Torpidity
> And boredom hold their lasting sway
> With unconcerned Stupidity;
> A land where old Obedience sits,
> Well-filled with Faith, devoid of Wits.

Lord Byron thought differently:

> There was a sound of revelry by night,
> And Belgium's capital had gathered then
> Her beauty and her chivalry, and bright
> The lamps shone o'er fair women and brave men.

First there should be a word about Belgium in general before proceeding to discuss the capital, because the city is impossible to grasp without the country. Indeed, Brussels is the country's principal problem, an unusual circumstance. Washington is not America's greatest problem, London is not Britain's greatest problem. The reason why Brussels is such a problem is the bitterly contested issue of language, to be described in detail later.

Suffice it to say now that the country is cut straight across into two divisions by a line as clearly marked as by a saw going through a plank—Flanders in the north, populated by the Flamands who speak Flemish, and Wallonia in the south, where the people are French-speaking. Brussels is bilingual, and so is the administration of the country at large. It even has two names —*Koninkrijk België* in Flemish, *Royaume de Belgique* in French. *Everything* has to have two names. I bought a tube of American shaving cream packaged for the Belgian market. The label says both *"Crème à Raser Sans Blaireau"* and *"Scheercreme Zonder Kwast."*

The Belgians are the most Latin of the German peoples, the

most Germanic of the Latin peoples—an old aphorism which puts the situation nicely.

The country lies athwart the principal route from the Rhine to the English Channel, which made it a pathway for conquerors from the earliest times. Gauls, Huns, Franks, marched and fought here. It has experienced intruders from Julius Caesar, who wrote that the Belgians were the most courageous of all the Gauls, to Kaiser Wilhelm and Adolf Hitler. It has lived variously under Roman, Frankish, Burgundian, Spanish, Austrian, French, and Dutch rule. Above all, it was trampled on and despoiled by nations who fought out their *own* quarrels on Belgian soil, naturally without asking Belgian permission, as several historians have pointed out. Marlborough fought here. So did Napoleon— everybody knows that Waterloo lies just outside Brussels. Once a few centuries ago Belgium was an independent republic—for two whole days.

Modern Belgium dates from the Congress of Vienna in 1815, when the great powers allocated it to the Netherlands. This arrangement, which took no account whatever of Belgium's own desires or interests, was largely a British device to keep the country from becoming part of France. The British, playing balance-of-power politics as always, or at least when they were sure they could tip the balance, wanted the Low Countries to be a neutral buffer across the Channel. After fifteen years the Belgians revolted against Dutch rule, and (1830) declared their independence. They have retained this ever since except for the two periods of German conquest and occupation in World Wars I (1914-18) and II (1940-45).

The British picked a pork-faced German princeling, Leopold of Saxe-Coburg-Gotha, who was one of Queen Victoria's uncles, as Belgium's first king in 1831. He had served in Napoleon's court, held a commission as a general in the Russian Army, and on one occasion was offered the throne of Greece. On its foundation Belgium was not considered to have a true national identity (any more than the King had), and the new monarch was called

"King of the Belgians," not "King of Belgium." Officially this usage persists to this day.[1]

Belgium may be synthetic, an artificial creation, but its cities are not. The country, which can easily be crossed by car in a day, contains a profusion of stalwart small cities—Antwerp, Bruges, Ghent, Ostend, Ypres, all in Flanders, and Liège (Luik in the Flemish language), Namur, Mons (Bergen), Charleroi, Bastogne in Wallonia. The celebrated university town, Louvain (Leuven) lies in Flemish Brabant near Brussels. These stout little cities, towns, and communes—about 2,500 in all—are the permanent, indestructible bases of Belgian history and culture. Each has a decided personality, and most have been wealthy from the earliest times—city-states, in effect, with a strong tradition of stubborn free will and independence. As one distinguished citizen said to me, "The communes are the heartbeat of our democracy."

Middle-classness is still a dominant note, accompanied by solid vitality, desire to make money, and optimism. The cabinet even includes a Minister for the Middle Classes—shopkeepers, artisans and the like—to protect the interests of these worthy citizens. Workers, too, are well protected, and social security payments are substantial, although not so high as in Paris. Caution, industriousness, and the mercantilist tradition are implanted in every element of the populace.

The country is profoundly Catholic, and its court is the last Catholic court in Europe. King Baudouin and Queen Fabiola de Mora y Aragon, his Spanish bride, are seldom seen; they are respected, but not particularly loved. Baudouin's father, Leopold, who was forced to leave the throne in 1951 after a double-faceted struggle too complex to go into here, still lives near Brussels, but is altogether out of the picture. There is no "Royal Question" now, but Brussels has the strange distinction of having two monarchs, a father and a son, one ex, one active. Court life is stiff, and a car passing the royal residence in Laeken, on the out-

[1] This form of nomenclature was also intended to indicate that the King did not have divine right like Louis XIV et al. Napoleon similarly called himself "Emperor of the French," not "Emperor of France."

skirts of the capital, is not permitted to stop. The king has a civil list of 36 million francs, a somewhat imposing sum, and other members of the Royal Family receive a substantial civil list.

Belgium has a good many men of considerable wealth, but there is no conspicuous aristocracy. The great fortunes rose out of copper (the Congo), finance, insurance, and, more recently, industry, although several lucrative industries have been long established. The big money is in the hands of a few, and most of the rich—not all—live unostentatiously. On the other side of the fence few of the poor—in Brussels at least—seem really poor. There is no industrial proletariat, and I was hard put to it to find a slum.

But other than merchants have labored in these flat neighborhoods. The roster of Flemish artists includes Hieronymus Bosch, the Van Eycks, Hans Memling, Pieter Brueghel, Anthony Van Dyke, and, above all, Peter Paul Rubens—a positively Florentine accumulation. Erasmus (born in Holland) lived here for many years; so did Victor Hugo. In other fields there were Mercator the mapmaker, and, much later, César Franck. The inventor of the saxaphone was a Belgian, and so was the discoverer of the Andes. Georges Simenon, the detective-story writer, was born in Liège, and the Parisian avant-garde writer Françoise Mallet-Joris was Belgian-born. The first railway on the European continent connected Brussels and Malines in 1836, and some fifty years later the first international telephone call in history was put through between Brussels and Paris. There have been three Belgian winners of the Nobel Peace Prize, but I never met anybody who knew much about them.

Brussels, a Burgher's Paradise

Brussels is shaped like a swollen heart. Being bilingual, the city has two names—Brussels in Flemish, Bruxelles in French; both derive from a Dutch phrase meaning "dwelling on a marsh." I thought it was one of the few important capitals I have ever seen without a river, but no—Brussels does have a river, the Senne. But it was boarded over some ninety years ago in order to eliminate the contagion supposed to emanate from its dank waters, and it still runs underground below the city streets until it is allowed to burst to the surface outside the urban area.

The Bruxellois are great builders, and the city has no fewer than 493 public buildings of allegedly historic interest, including several monstrous eyesores. The population of the metropolitan area is around 1.1 million—roughly one-ninth of that of the country—and remains stable. In fact, the number of people is actually diminishing a bit, as more and more citizens move out to the suburbs. There is no population explosion here.

This is an old city; it was almost a thousand years old when the Dutch, first cousins of the Belgians, bought Manhattan. The foundation date is put at A.D. 706, and Charlemagne made a visit here in 804. In 1430 it became the capital of the merged duchies of Brabant and Burgundy, reached a population of forty thousand, and, like Malines nearby, grew famous for its cloth, its weavers. But it was a mere "hunting lodge"; it did not have the prestige, weight, or wealth of Bruges, Ghent, Liège. Then it became for a time the capital of the Spanish Netherlands, because it was a bridge between the French and Dutch. So many kings have ruled here, at least titularly, that Brussels is sometimes called the Citta Regale.

Heavy industry never touched Brussels, and I never saw a factory chimney of any size. But the city became the financial center, the indispensable banking outlet, for industrial communities outside like Ghent and Antwerp; industry was *managed* here, and it became the Wall Street of the nation. One short street today contains the offices of more than twenty insurance companies. One mammoth bank, the Société Générale, is supposed to control today not less than 40 percent of *all* Belgian industry, as it once controlled 70 percent of the wealth of the whole Congo.

Lace and china are the surviving local crafts—one can scarcely call them industries. Lace goes back to the sixteenth century or earlier, and was originally dependent on pauper labor; then women of the bourgeois class became the lacemakers, and there were ten thousand of them in Brussels in the 1700's. Only a handful work at lace today. Brussels lace can be very expensive: $20 for a handkerchief, several thousand dollars for an elaborate bridal veil. One young woman of my acquaintance, who prefers Maltese lace to Belgian, says that the latter is "neurotic" and "hysterical," made by women who manufacture it with the same gesture that they would use picking nits out of their hair.

What, at first glance, is by far the most striking thing about Brussels is the proliferation of tall new glass-and-steel office buildings, an impression which is particularly marked if the visitor arrives from such a predominantly flat city as Paris. Brussels has more skyscrapers than any capital in Europe. From the roof garden of our hotel we counted seventeen. The highest office building in Europe (thirty-eight stories), the Tour de Midi, housing the National Pensions Administration, stands here. Even so, Brussels is not a New York or a São Paulo by any means. One thing I noticed is that traffic arrows pointing to the center of the city are marked:

$$\text{CENTR}\begin{cases} \text{E} \\ \\ \text{UM} \end{cases}$$

The economically-minded Belgians thus spell out the word "Center" in both French and Flemish, as is obligatory, without the expense of reprinting the entire word.

Next, I think, the visitor will notice greenery and open spaces. There are parks all over the place. The metropolitan area shelves off abruptly into a green plain on one side, and on the other into a heavily wooded area, the Bois de la Cambre, which is part of the Forêt de Soignies, a real forest, not a park. Belgians are mad for greenery, and thousands pack off every weekend for the Ardennes. On roads nearby, and also en route to Ghent and Bruges, there are multitudes of bicycles, and the visitor will see languid fields of hops, a few poppies, a windmill or two, and silver power lines dancing over pigs as big as calves.

Brussels itself is, however, pretty ugly, and, in spite of the new skyscrapers, looks somewhat unkempt. The streets are not, as in Paris, swept clean every day. The trams are cream-colored, and the taxis (which do not cruise) are orange. New construction is wrecking one neighborhood after another, which means that the quality of the old city is steadily being eroded. There is little sentimentality about antique buildings.

One distinction is an inner ring of boulevards, modernized for the 1958 World's Fair, which follow the semicircular line made by ramparts in the fourteenth century. There is nothing like a bright snappy World's Fair to put a city on its toes. A long straight street, which changes its name several times but is best known as the Boulevard Adolphe Max, was named for an illustrious burgomaster who held office for forty years. One big Brussels landowner is, of all people, said to be the Emir of Kuwait, whose holdings are supposed to be worth $40 million. Many shops nestle in arcades, or hug the rims of leafy squares; this is convenient for shoppers, because it rains 220 days a year. The antique shops near the Place du Sablon are somewhat dull. There seems to be no housing shortage, and rents are not particularly high. Foreigners arriving to work with NATO or the Common Market do not, as a rule, find it difficult to find pleasant accommodations, and servants are still procurable at a modest wage—a good Flemish cook isn't likely to earn more than $80 to $120 a month, a cleaning woman 60 to 80 cents an hour.

Brussels is a great town for markets. There are nine open-air food markets, an antique market, a flea market, bird and flower markets in the central square on Sunday, and a flourishing bicycle market. On Wednesdays, as I heard the scene described, the countrymen flood into the metropolis to sell their produce, and the town takes on a jovial, rotund aspect, like something painted by Brueghel at his jolliest: the restaurants are thronged, red-faced burghers full of beer stagger through narrow medieval streets, children cavort, boulevard cafés echo with hoarse laughter, and prostitutes, stout fresh-looking country girls for the most part, do a land-office trade. I have never seen a city where so many women (virtuous) stand alone on street corners. I don't know why. And every neighborhood has middle-aged women sitting in third- or fourth-story windows staring passively, almost without movement or expression, for hour after hour, at whatever takes their interest on the street.

Brussels is a city not merely of small shops but of small businesses. An astonishing number of these have, however, a very short life indeed, in spite of present-day prosperity. They are insects, which die at night. No fewer than 35,000 new businesses were set up in Brussels last year, and 34,000 failed. This is strange—write it down as another Brussels paradox. Bigger companies are, on the other hand, flourishing. One development is the "invasion" of Belgium by any number of American companies establishing big operations here—General Foods, Corning Glass, Cessna, Sears, Pfizer. They are welcomed by the Belgians —no government permission is required for setting up a company in Belgium as a rule—and they operate tax-free.[2]

Rational the Bruxellois seem to be, but look at their driving regulations. Accidents are horrifyingly frequent. We saw the grisly aftermath of two on successive days. The carcass of one car looked like a crushed tin can. Cops in bright yellow-orange slickers (to make themselves so conspicuous that *they* will not be run down?) measure out the scene carefully, leaving every fragment of wrecked car in place, setting up markers, tracing skids, taking photographs. Accidents are so common and so damaging that cynics say that carnage on the roads is what keeps the population down. Another crack is that the authorities are renovating the streets, but not the drivers. The amazing fact remains that, until new legislation was passed which is not yet fully in effect, anybody over eighteen can get a driving license *without passing a test*, and licenses were not even obligatory until recently. Yet, in other dimensions in this field, the law is severe. If, when

[2] *Time,* February 2, 1968. About $780 million have been invested in Belgium in the past ten years by American companies.

parked, you do not lock all your doors, the fine is 100 francs per door.

Belgians, it would seem, do not have much respect for the law in the abstract, probably because they have had to live under so many foreign regimes, but they are essentially law-abiding because this makes good sense. The Brussels commune (a term to be defined later) has about three thousand police, and only about fifteen thousand crimes and offenses were reported in the whole of 1966, fewer than in an average month in a big American city. These included one plot against the Royal Family, 8 false claims to nobility, 65 swindles, 101 attempts at suicide, and exactly 3 murders, of which only one was considered to be a true murder, because it was premeditated. The other two were classified as "assassinations."

One morning my wife and I went to see Léon Cappuyns, the Vice Governor of Brabant. The major task of this able, agreeable man is to develop long-range plans for the community in such realms as housing, education, commuter services, traffic, and the like. Forty-seven, devoted to the public good, sandy-haired, a lawyer by profession, Mr. Cappuyns learned English during the war when he was a Resistance fighter and took part in the underground movement to smuggle RAF airmen who had been shot down on Belgian territory out to safety. Mr. Cappuyns is a Fleming and, somewhat unusually for Flamands, a Protestant— also hard-working, modest, and articulate.

Housing, he told us, is a subject full of pitfalls because Belgians do not like to live in huge apartment blocks; they are builders who want small dwellings of their own. Traffic is not the agony that it is in Paris or Rome, but can be troublesome. Parking is a headache. "It is not fair for a private car to occupy fifteen square meters of public property for long hours." Traffic is now being expedited along the circular boulevards by means of a series of artfully constructed tunnels. "But we are always a tunnel behind," Mr. Cappuyns said.

✤

Brussels's classic sight is, of course, the Grand'Place (written just like that in French—no space after the apostrophe), which is indeed one of the most arresting sights of Europe, precious to the connoisseur. It is not grand at all—measuring no more than 350 feet by 200—but it has a unique sheen and sparkle, with its principal buildings delicately scrolled with gold. The Grand'-Place dates back to the fourteenth century, and is the womb of Belgium. French invaders destroyed most of it by artillery fire in the time of Louis XIV, but it was lovingly, diligently restored, and, centuries later in 1960, it was formally named "a sacred island," meaning that never, under any circumstances, may its buildings be tampered with or torn down to make room for anything else, not even a new skyscraper. This is one case where culture outrides commerce.

On one side of this pretty jewel box, which is illuminated gracefully at night, is the Hôtel de Ville, with its gilded spire, which dates from 1402; on the other is the King's House, better known as Bread Market Hall (Broodhuis). Thirty-nine other guild houses sit elsewhere on this square, representing such ancient Brussels corporations as the archers, barrow makers, millers. Then there are several tiny but admirable restaurants, jewels within the jewel box, and two cafés. Byron lived here once, and Jean Cocteau called it a "theater unique in the world." But dirty-looking beatniks throng its pavements now.

Not far away rises a hideous structure known as the Palace of Justice, the largest building put up anywhere in Europe during the nineteenth century; it covers 26,000 square meters, more than St. Peter's in Rome. It is probably also the ugliest building on earth, although Rodin did some of its decoration. Another sight is the Atomium, the only surviving ornament of the 1958 World's Fair. A dramatic aluminum structure, it demonstrates how atoms, magnified 200,000 times, form a molecule. Then, too, Brussels has the celebrated fountain known as Manneken Pis, whose official name is Little Julian. This small statue of a boy playfully making water dates from 1619, and Julian is affectionately known as the "first citizen" of Brussels. Delegations from foreign countries intermittently visit Manneken Pis and ceremoniously present him with costumes designed to fit his exposed condition—even kilts, Lord Fauntleroy pants, and a Boy Scout uniform; these—there are several hundred—are put away in a museum for permanent display. Vandals attacked Manneken Pis in 1965, breaking him in half. At once the outraged town authorities ordered him to be put together again.[3]

Still another sight worth a visit is an odd little collection, the Museum of Musical Instruments, containing amongst much else an arrangement of bells tapped by hammers which was invented by Benjamin Franklin. The legend is that this contraption was barred from use in the United States out of fear that it might cause deafness in the operator; Franklin brought it with him to Paris, and eventually it found its way to Belgium.

Nineteen Cities, Not Just One

The most unusual and distinctive thing about Brussels is that it is not one city, but nineteen different cities or communes, all separate and autonomous. For the sake of convenience they are bound together as an "agglomeration," but all nineteen are independent entities. No such thing as "Brussels," meaning the whole capital, exists officially. The name is used on maps, in the postal service, in foreign contacts, and so on, to indicate the metropolis as a whole, but the city is not a unified organism, not a single administrative or political unit. Correctly, "Brussels" is merely the central commune among eighteen others of equal rank.

The Brussels commune contains, however, most of the big banks, the railway stations, hotels, Parliament, and the ministries and administrative offices for the country at large. Its population is about 170,000 out of 1,100,000 for the whole of the city. It is oddly shaped, with an appendix hanging down to the southeast. Among the other communes are Anderlecht, a commercial quar-

---

[3] The Michelin Benelux Guide says that Manneken Pis symbolizes *"la goguenardise et la verdure du bon peuple brabançon."*

ter, Évère near the airport, Laeken with its royal preserves, Schaerbeek, Saint-Gilles, and Jette-St. Pierre. Most have a distinct, separate history and development, with frontiers dating back to the seventeenth century or earlier. Each has its own burgomaster and local administration, each is separately enfranchised as a city, and in theory none yields to another.

There is no precedence among the nineteen mayors, but an informal cooperative apparatus has arisen; for instance, all the burgomasters meet regularly in the Town Hall under the chairmanship of the Mayor of the Brussels commune—although this official does not hold the chairmanship as a matter of right but merely as a convenience. The communes guard their institutional privileges with the keenest zeal, which can lead to paralyzing administrative confusions. For instance, there is no unified police force, no central police station—not even a single number to call on the telephone in an emergency. Until recently it was possible, at least in theory, for a criminal to hop across the street from one commune to another and claim immunity to arrest even if the police were hot on his heels, on the ground that they had no jurisdiction in the area where he had taken refuge. Laws vary from commune to commune. A corner property may be located in two or even three different communes, each with its own tax structure; a man may pay land taxes in one commune, other taxes in another. Statistics on a pan-Brussels basis are almost impossible to obtain. Traffic regulations differ street by street. Then consider fire. There are six different fire departments (several of the smaller communes share their forces), and their equipment does not necessarily match. In May, 1967, the Innovation department store in the Brussels commune was destroyed in a disastrous fire; hundreds died, and bodies were still being pulled from the ruins months later. Brussels called on its sister fire departments to help, but one reason why the fire was so difficult to control was that hoses from one commune did not fit others.

Normally the visitor to Brussels has no conception of these confusions because the communal frontiers are unmarked and anybody may pass freely from one to another. There are no signs, like "City of Westminster" in London, to define where a person is. The irrationality of this system—in supposedly "rational" Belgium!—has provoked wide and lively demands for reform. One reason why this is not forthcoming is, of all things, the fact that the Germans, during their occupation in World War II, did temporarily abolish this medievalism and consolidated the city into a single administrative organism, called Grossbrüssel. Immediately after liberation the Belgians restored their own way of doing things, and they resist going back to anything that smacks of Germanism.

The burgomaster of Brussels commune is Lucien Cooremans, a brisk vigorous man, about fifty, short in stature, who is also the chief of police. He received us early in the morning in the Hôtel de Ville, offered us our choice of champagne or whiskey, and smoked a bristly little cigar with a gesture of cheerful impatience. The Mayor said that the commune system was cumbersome and antiquated. His "Brussels" comprises one-sixth of the city in area, but he has only one-nineteenth of the authority. These are any number of anomalies. One commune, Ixelles, is cut in two by the Brussels "appendix." The council of nineteen burgomasters, of which he is chairman, has no power of decision. Each commune has its own fiscal powers, to which it holds adhesively. Cooperation is possible in some fields, but in financial matters, no. Integration, Mayor Cooremans feels, is so necessary that it is bound to come.

Each commune is in theory run by councilors elected by the people for six-year terms; Brussels has thirty-nine. The councilors have no executive authority, and, as in Paris and London, receive no salary except a token *jeton de présence*. The burgomasters of each commune, who also serve six years, are appointed directly by the King from among the councilors; reappointment for other terms may follow, as has happened in the case of Mr. Cooremans. The Belgian monarch thus has a distinct power in municipal affairs, although he is of course constitutionally advised.

I asked a non-Belgian why the nineteen communes could not

be united into one. Reply: "That would leave eighteen burgo-masters without jobs."

Education, the Arts, and the Table

Intellectual and artistic expression is not pronounced in Brussels, partly because Paris, which is only three hours away by train, an hour by air, sucks off most of the incipient French-speaking talent. There are two state universities in the country (Ghent, speaking Flemish, and Liège, speaking French), and two private institutions, the University of Louvain (Catholic) and the Free University of Brussels. The latter has 6,934 French-speaking students, 747 Flemish. It is supported in part by the state, and has strong Protestant—also left-wing—elements. Instruction takes place in both languages, which makes for confusion. It is difficult to express a culture simultaneously in two languages. Ugly riots based on the language issue occurred in Louvain in 1968, so severe that they provoked the resignation of the national government. Louvain itself passed into "a state of siege."

TV is indifferent, with one Flemish channel, one French. You can choose your language version of such American cultural imports as *Hollywood Panorama* and *Dr. Kildare*. The radio is better. Some Americans here object hotly to its political commentaries, however, because it takes on occasion a vivid anti-American line, especially about Vietnam. The Belgian Government, which is pro-American, can do little to influence this because the Belgian radio is set up under an independent authority rather like the BBC in England. There are three Flemish stations, three French. The French stations are more left-wing.

Brussels has a well-established opera, the Monnaie, one of the oldest in Europe; it was founded in 1700, and Richard Wagner once conducted here. The city has a strong and active musical tradition. Musical scholarships and awards in the name of the late Queen Elizabeth, the widow of King Albert, who was an accomplished musician herself, draw contestants from all over the world.[4] One cherished figure of the recent past is the violinist Eugène Ysaye. The city has a sound symphony orchestra, and its ballet company, directed by Maurice Bejart, who was much influenced by Martha Graham, is one of the most advanced in Europe. The theater (French) is not much. There is a Flemish national theater as well.

Good painters are few, and the best go to France. One artist with a world-wide reputation is René Magritte.[5] A group called the Expressionist Flamands made a stir some years ago, but has petered out. As to writers, most nationalist Flamands do not know French well enough to write in French, or, if they do, they insist on writing in Flemish anyway, which means that they have to be translated into French or some other world language in order to become known abroad. It is interesting to reflect on what would be the reputations of the two greatest modern Belgian writers, Émile Verhaeren and Maurice Maeterlinck, if they lived today under the full pressure of the language crisis. Both would probably have chosen to write in Flemish.

Brussels loves to eat. Belgium has been called the civilization of fried potatoes, and indeed *pommes frites* are served at almost every meal and sold on the streets like popcorn. Several places— like the souvenir stalls on the battlefield of Waterloo—fairly reek of them. But the *haute cuisine* is superb—if somewhat rich—and, although Brussels is not a very big city, I would say that it has at least three restaurants that are among the dozen best in all Europe. Most of the really good restaurants are tiny (one has only eight tables), and several have been in the same family for generations; they are intensely serious, expensive, elegant in a *fin de siècle* sort of way, and absolutely classic in their procedures. I would cross the Atlantic on five minutes' notice to taste again the

[4] Iccidentally, it was this lady, rather than Winston Churchill, who invented the phrase "Iron Curtain." German born, she announced in 1914 that she had renounced her Germanism forever when the Kaiser's forces, by invading Belgium, erected an "iron curtain" between herself and her native country. Queen Elizabeth, the daughter of an oculist, was a person of varied talents—a nurse, doctor, Egyptologist, poet, and playwright. New York *Herald Tribune*, November 24, 1965.

[5] Mr. Magritte died after this chapter was written.

delicate consommé with cherries at the Épaule de Mouton, the filet of sole with a mushroom and truffle sauce at La Couronne, and practically anything at Comme Chez Soi. One new hotel, the Hilton, has a room which serves beef flown every day from the United States—"air-fresh beef"; here stout Bruxellois stuff themselves till they turn black in the face. But not only are the leading hotels and restaurants good; practically all restaurants are good—the general level is extraordinarily high. One street near the center of town is devoted exclusively to restaurants. Incidentally, I never saw a Brussels sprout in Brussels.

Beer is the national drink, and Belgians are reputed to be the third largest beer consumers per capita in the world, after Czechs and Germans. At least thirty different types of beer are known, including one, called Kriek, which is flavored with cherries and another, Gueuze, which is brewed from, I believe, wheat. A sweet beer is called Faro. I liked best a standard brew called Stella Artois. Astonishingly enough, Belgium has severe restrictions about hard liquor. The day I arrived I walked into a bar, and, to my surprise, was asked by the manager to sign a book. I thought appreciatively that this was a gesture of goodwill, but no—it was to make it appear that the bar was a "club," so that whiskey and cocktails could be served to me, since I was now a "member." Then one evening I sauntered up a boulevard, sat down at a sidewalk café, and ordered a cognac. I did not get it. Spirits are forbidden except in a "club." And even the most fashionable clubs are obliged to display a conspicuous certificate entitled "Repression of Drunkenness," which sets forth the rules. Of 3,074 arrests in Brussels in 1966, more than a thousand were for public drunkenness. Semiprohibition came into force many years ago mainly as a matter of public health and to protect the worker's cash income. Some people say, not quite seriously, that it is maintained today largely by the influence of the beer lobby to deprive people of rival alcoholic fluids.

Brussels is not exactly what you would call a gay city. A Moulin Rouge exists, even a Folies Bergère, feeble imitations of

Paris, but this is not a place for night owls. On one Saturday night I found that the bars in two leading hotels were shut at half-past ten.

The Fanged Issue

Underlying almost everything in Brussels is the drastic pressure of the language problem. The issues that perplex most great cities —population pressure, housing, slums, smog, juvenile delinquency, segregation, crime, education, politics, municipal welfare, the state of business—do not exist on a critical level here. But the language problem is a bitter divisive factor, which seems to be becoming more insoluble as time goes on. Belgium, we know, is divided into mutually exclusive linguistic strips. Flanders in the north, with its tenacious Teuton-descended Flamands, has 5,265,000 people; their written language is identical with Dutch, and they speak a variety of Dutch. Wallonia in the south has 3,900,000 French-speakers, sometimes called "Francophones," and has close French affiliations. Of course there are thousands of mixed families. It is impossible to tell who is a Flamand, who is a Walloon, by the name alone.

Plenty of Flemings speak French, but few French speak Flemish. The Flemings cross the linguistic barrier much more readily than the French. By and large a person speaking French is pretty much isolated in rural Flanders, but this is not so much true of a Flamand in Wallonia. As to the French, they simply will not bother to learn Flemish, which they consider to be a minor language totally alien to them. Interestingly, almost all educated Belgians speak English well, partly because they prefer to learn English rather than their own second language.

Brabant, the only province out of nine which does not touch the sea or foreign soil, is the core of the struggle. Its capital, Brussels, is, as I have mentioned, bilingual, but its northern half is Flemish-speaking, the southern French-speaking. To say this is to oversimplify an exceedingly complex situation. Though largely French in speech, Brussels is Flemish by most other criteria; the figures usually given are that it is 80 percent Flemish by blood, but 80 percent French by language; in a word, it is a Flemish city that has become Frenchified. Moreover, Brussels lies just north of the language frontier and is thus embedded in *Flemish* territory; it is a French-speaking enclave within a Flemish domain, or, as some Flamand nationalists say, a Flemish city "captured" by the French. One result of this is that both the Flemish and French communities outside Brussels tend to dislike the capital. The Flamands deplore it when a Flemish-speaking person moves into Brussels from the countryside, on the assumption that he may become "lost" because he has entered a bilingual society. The French, on their side, seek to increase their influence; the Flamand nationalists say, indeed, that the French intend "to take Brussels over." Above all the Flemings resent the contemporary exodus of prosperous French into the suburbs, because this means the extension of French influence into traditionally Flamand areas. A fight over education is part of this. Instruction in both languages is compulsory in the nineteen communes of the Brussels amalgamation, but in peripheral communes French may be taught at the public expense only to students whose *mother* tongue is French.

More than mere language enters into this painful situation, because what really counts is caste—economic and social bias. The problem is basically social, not linguistic. After independence in 1830 the French-speakers became the upper class—top dog. They were the aristocrats, the privileged. They ran industry, the big banks, and the government. They were educated and exclusive. The Flamands were poor, lower-class, and almost totally excluded from "society." They were uneducated peasants, and were looked down upon as such. It was virtually impossible for the Flamands to rise to a big position in the bureaucracy, business, or the professions, if only because they had little opportunity for education and did not speak French.

Until comparatively recent times a Flemish factory worker could be fired by a French-speaking foreman without being able to understand what he was being fired for. An army recruit could be severely punished for an infraction of discipline without knowing what his commander was saying. A Flemish boy picked up for an offense might not be able to understand one word of the proceedings of the French-speaking court where he would go on trial, or even the meaning of the sentence the judge imposed.

A sophisticated member of the nobility said to me ruefully, "We in our family spoke nothing but Flemish until a couple of hundred years ago. Then we turned to French. Now we are obliged to learn Flemish again. We thought that Flemish was a language spoken only to servants and farmers, but now the farmers are cabinet ministers."

The renascence of the Flemish began seriously after World War II. Their birth rate is high, and they became the biggest bloc in the country; they grew more prosperous as well. Their economy advanced, while that of the Walloons declined. The Flamands thirsted for revenge after years of discrimination against them, and demanded equality of treatment. The issue became intermeshed with politics at every level, and led to profound instability. Seemingly the most stolid of states, Belgium has been close to civil war twice within the last twenty years. Another factor is religion. The country is overwhelmingly Catholic, but the Walloons are more "laïque" than the Flamands. So the spectacle arose of people taking sides on religion as well as language—holier-than-thou Flamands versus worldlier-than-thou Walloons. Then came the formidable crisis over the "Royal Question" in 1951, which split the country further. Profoundly difficult labor troubles followed, with a period of distracting strikes. Flemish labor is much more conservative than that of Wallonia, which is the heart of Belgian socialism.

There are three more or less evenly balanced political parties in Belgium, and government usually takes the form of an uneasy coalition. The Social Christians are the Catholic party, and their principal strength is in Flanders. Leadership of the Liberals (strong conservatives) is mixed. The socialists are mainly French-speaking and Walloon. The prime ministership more or less alternated for a time between Flamand and Walloon, and the tradition is that the Prime Minister must be bilingual, which sometimes means that he is equally bad in both languages. The present incumbent is a Bruxellois; his deputy is a Flamand, but one who speaks French; the Foreign Minister is a Walloon who speaks Flemish. Several lesser ministers are not bilingual, and an interpreter has to be present at cabinet meetings, a circumstance unparalleled in the world.

A fantastic amount of duplication of effort and expense is made necessary by this language issue. All official documents and correspondence have to be in two languages. The French resent it bitterly that their children are obliged to learn a "dead" language, of no use to them whatsoever in later life unless they live in Flanders or the Netherlands. But the Flamands are adamant. Recently they were strong enough to push through a law which provides that, by the end of next year, 50 percent of *all* government jobs must be held by Flemish-speakers, from ministerial chiefs to sweepers in the post office. Moreover, all important officials must now pass an oral and written examination to prove adequate knowledge of Flemish. Even diplomats on active duty abroad, no matter what their seniority, are obliged to submit to these examinations. Recently a fifty-year-old Walloon ambassador to a French-speaking country was called home to take his Flemish exam, and flunked it. He was then replaced by a Flemish civil servant who had to learn French in order to take over the post —a crazy episode.

Not once did I talk to a Belgian official without the language problem coming up. I have mentioned our visit to Léon Cappuyns, the Vice Governor of Brabant. As soon as he had sketched his problems in urban planning he launched eloquently into a discussion of the language schism. Flemish gave way to French because, in the old days, everybody civilized in Europe *had* to learn French; even Frederick the Great "fought his wars in French." Then, when Belgium became attached to the Netherlands in 1815, resentment against the Dutch suzerainty further increased pro-French sentiment. But now the center of gravity has shifted, and Flemish nationalism is resurgent. One reason for this, odd as it may seem, is the world-wide change from coal to oil; the coal fields in the French-speaking areas, which had been the basis of Belgian economy, lost their importance, and Antwerp, the great Flemish port, became more prosperous. Industry moved toward the sea; so did foreign investment. It is now Flanders which is rich, not Wallonia, a reversal of the previous position, and thus the Flamands have become more capable of self-expression.

Why cannot bilinguality be installed all over the country?

Because the French will not learn Flemish.

Solution?

The Flemings must be accepted on equal terms.

I talked one day with Paul-Henri Spaak, now sixty-nine, the grand old man of Belgium, a roly-poly astute man with a winning smile. An "Atlanticist," a world citizen, Spaak led the Belgian delegation to the San Francisco Peace Conference in 1945, served as first President of the UN Assembly in 1946-47, helped to create Benelux, was the first President of the European Assembly in Strasbourg, became a leading figure in the struggle to achieve the Common Market, and was head of NATO from 1956 to 1961— an impressive record.

A man of the moderate left, Spaak was the first socialist Prime Minister in Belgian history, and served as Foreign Minister for roughly twenty years out of the last thirty. He is out of office now. He does not speak Flemish. Mr. Spaak thinks that the root of the language problem is democratization. Until recently the upper classes—to repeat—spoke one language, the lower classes another, and this gave an acute inferiority complex to the Flamands. They were ordered about in a "foreign" language. Now today they are in a position to set themselves against the French Establishment, and this is what makes the crisis. Intermarriage is no solution because when it occurs it is almost always the Flamand who is assimilated, not the Walloon.

Brussels International

Belgium has been a leader in movements for European rapprochement and consolidation for many years, and more than 250 international organizations have their headquarters in Brussels. As far back as 1921 the government took the lead in setting up an economic rapprochement with Luxembourg, the first such arrangement between European sovereign states in history. Thirty years later came the formation of Benelux, an economic union comprising Belgium, the Netherlands, and Luxembourg, which was a major step in the movement for European integration. The great powers opposed it at first, and then gave way.

In the 1950's came the formation of two new important agencies which carried the Benelux conception further—the European Coal and Steel Community (ECSC) and Euratom, the European Atomic Energy Community. Each had six partners—France, Italy, West Germany, and the three Benelux states. This was an experiment on a new, bold, and altogether unprecedented scale. The ECSC set up its headquarters in Luxembourg; Brussels became the "capital" of Euratom. Then, after 1957, came the creation of something even broader and more seminal—the European Economic Community (EEC) or Common Market. Jean Monnet, the French man of affairs, and Paul-Henri Spaak were its principal architects. Establishment of the Common Market, the ultimate aim of which was the abolition of tariff barriers among its members and associates, has had profound international repercussions, as everybody knows. A European Assembly was set up in Strasbourg and a High Court established in Luxembourg—all of this as part of a scheme designed to create in the long run nothing less than a federal Western Europe based on economic unity, with political unity to follow.

Brussels became capital of the Common Market, after a lively tussle with Rome and Paris, both of which wanted it. Headquarters were established on a street named, interestingly enough, the

Avenue de la Joyeuse Entrée.[6] Dr. Walter Hallstein, a German civil servant close to Chancellor Adenauer, became its first chairman, and the staff, a handful at first, grew to seven thousand. Euratom has about fifteen hundred. These "Eurocrats" give a special coloration to the life of Brussels today. They have their own license plates (a blue "EUR" followed by a number) on automobiles as well as certain extraterritorial privileges. They make Brussels more cosmopolitan than it has ever been before, and add to its prestige. The cream of international diplomacy flows incessantly through Brussels, together with hundreds of officials, experts, and technicians in various fields and of various nationalities.

In July, 1967, came a further large forward step—the merger of the Coal and Steel Community, Euratom, and the Common Market into a single organism, under the name "European Communities" and with Brussels as its capital. Hallstein withdrew as chairman of the over-all fused organization, and was succeeded by a Belgian named Jean Rey, who symbolizes the "New Men" of Europe. Hallstein was somewhat arid and overgiven to show and protocol, but Rey is more relaxed—a smiling as well as sophisticated character, vigorous, outgoing. Rey's remote origins are Italian, but his family has been associated with Liège for generations. His great-great-great grandfather was its first mayor after liberation from the Dutch in 1830, and his grandfather, a congressman, held the same position. Rey is a Protestant—indeed, Belgium's "Mister Protestant." After the war he served briefly as a professor of international law at Georgetown University in Washington, D.C.

The new European Communities are to be housed in an enormous star-shaped building on the Rond-Point Robert Schuman, named for the French statesman who was a pioneer of the Coal and Steel Community. All the upper bureaucracy of the organization have an unshakable *idée fixe*—the eventual creation of a United States of Europe. Meantime NATO, the North Atlantic

[6] The name dates from a charter giving Brabant a democratic constitution in 1356.

Treaty Organization, also moved to Brussels out of righteous indignation when General de Gaulle tossed SHAPE (Supreme Headquarters Allied Powers Europe), its military arm, out of France. As a result Brussels is one of the few world capitals with three different sets of foreign ambassadors. There are 92 embassies accredited to the Kingdom of Belgium itself, 71 to the European Communities, and 15 to NATO.

SHAPE set itself up in Casteau, a dreary Belgian village thirty miles from Brussels, under the command of General Lyman L. Lemnitzer, a former United States Chief of Staff. The Belgians put it out to pasture in the sticks, much to Lemnitzer's displeasure, because they did not want a big military installation in their capital. But NATO, the superior political instrumentality, was welcomed and was allowed to set up headquarters within Brussels itself. Provisional headquarters are being built at Évère, in an outlying district where a deserted airfield was available. About two thousand men and women comprise the present NATO secretariat in Brussels, not including the delegations of the fifteen member nations. The Belgians want the site back eventually, and so another large NATO installation, to be permanent, is to be built at Heysel, the seat of the 1958 World's Fair—cost $20 million. Apparently the Belgians assume that NATO will stick.

Thus, housing NATO and the European Communities, rich prizes both, with SHAPE close by, Brussels has become the capital of the Common Market and its associated bodies as well as the headquarters of the Atlantic Alliance and the Western defense system, comprising an arc all the way from Norway to Turkey. Here in Brussels are two vital keys to Europe's future—if it has one. Of course the paradox is striking that Belgium is severely divided within itself. Perhaps the very fact that it is so rent and riven gives a particular stimulus here to the ideal of a united peaceful Europe.

6. ENTERTAINMENT IN
HAMBURG AND VIENNA

About ten minutes after arriving in Hamburg I began to reflect on what makes a good hotel. Most good hotels in Europe are, I dare to think, substantially better than those analagous in the United States. Take the Vier Jahreszeiten here. Like most good European hotels it is, first of all, quiet; the walls are not made of Kleenex, and there is plenty of that delicious commodity known as wasted space. Ceilings are high, corridors are wide. Even single rooms have a small foyer, to act as insulation from the world outside. The furniture is gracious, and the beds, with triple mattresses, are comfortable. A large silken puff, or *plumot,* supplements the blankets, but this, for restless sleepers, is sometimes difficult to anchor down. The bathrooms are luxurious, although they may carry a note of the arcane, an element of mystery, in the best continental style. The toilet flushes by means of a rubber pedal on the floor.

Certain deprivations have to be conceded—there are no facial tissues, no pads or pencils by the telephone, no writing paper, and no morning newspaper. All services are, however, rendered almost instantaneously by the push of a button. What, indeed, makes good European hotels really good is, as goes almost without saying, the swiftness of the service, which is impeccably deft and polite as well. Then, too, there is the institution of the concierge, who is much more than a mere bell captain or keeper of keys, and who, if you know how to handle him, becomes the visitor's essential link to the life of the city, his umbilical cord as well as avuncular counselor.

The lobby in a really good European hotel, like the Vier Jahreszeiten, will not be disfigured by a thicket of raucous shops, nor by tangled crowds of men wearing badges to proclaim their identity. A few discreet *vitrines* may sound a note of elegance. An item not so pleasant is that the men's room in many German hotels contains an ugly porcelain contrivance built waist high into the wall—a vomitorium for drunks or guests who have dined or drunk too well. One detail more—in most really good European hotels, like Claridge's in London or the Gritti in Venice, you do not have to sign bills for drinks or room service; this is not merely a convenience, but a symbol of trust between the management and the visitor.

I arrived in Hamburg on a night flight from New York via Lufthansa, and dinner was the best meal I have ever had on an airplane. There were some fish balls, I think made of eel, a celebrated Hamburg dish, to begin with, and then a filet of rare beef, handsome as a sunrise, sliced individually by a chef in a tall white cap treading down the aisle. The dessert was a cake soaked in rum and cream and a basket of complex, festive chocolates. As for drinks, we started off with Steinhäger, a colorless spirit new to me. Lufthansa has adopted the sensible device of putting a buttonhole in your napkin, so that it can be fastened to your coat. I wonder if other airlines do this. I never happened to notice it before.

At the check-in counter when we arrived early in the morning, I encountered a splendid little example of the Teutonic mind on submitting my tickets to confirm my further passage to Vienna. A brisk young woman examined them and then said, passing them back to me as if they were something unclean and she was washing her hands of them, "This is the wrong day of the month. You are not here." I was too startled to reply. She repeated sternly, but in an entirely unconcerned manner, "There has been some mistake. You are not here!" She dismissed me, refused to deal with my tickets, and waved me grandly on.

I convinced her somehow that I was not a phantom and *was* there, and, escaping with a deep breath, took a taxi to the hotel. Hamburg is a true Northern city, descended from the Hanseatic League, and it smacks of the sea; it has always been a free state within the German configuration, and has a spirit of robust self-will. I have seldom seen a great city so clean, and it pullulates with flowers. The atmosphere is, however, strictly utilitarian and commercial. Hamburg, with a population close to two million, is the fourth largest seaport in the world, although it lies sixty miles up the Elbe; its prodigious docks are its greatest sight. Two pretty lakes, on one of which the Vier Jahreszeiten faces, dominate the center of the town on the Alster, a derivative of the Elbe. The shops are neat, copious, and well attended. A person could satisfy practically all the needs of life in the railway station, which is full of all manner of shops and booths. The theater is both lively and distinguished, and the opera has a monumental reputation. It played *Aïda, The Rake's Progress* (in English), and two Wagners while I was there. I went presently to the art museum. Here are Douanier Rousseau's well-known "Eve," depicting a dark, sharp-nosed Eve, extremely Semitic-looking, and what I think is the finest Manet I have ever seen—the representation of a gross and cynical man of the world inspecting the person of his mistress (or maybe wife) as she prepares her toilette—also good examples of modern German and Central European art.

When I finished dinner that evening and reached for a cigarette, the waiter said officiously, "Do not smoke. You must digest your dinner first"—another Teutonic touch.

Hamburgers are, it might be added (as every writer on the city has noted), unknown in Hamburg.

Then of course I went to the Reeperbahn, the red-light district in Saint Pauli near the river. *Mein Gott!* Here, indeed, Hamburg, despite its bourgeois quality, lives up to its reputation as a city with the maximum of vice and the minimum of temptation. One short street, cut off at each end by barriers, houses the official prostitutes. These, of all ages, sit behind scantily curtained windows and display their blowzy charms like ghosts silhouetted against dim lights. Throngs of men swarm up and down the street picking and choosing; there comes a rap on a window, and the woman leans out and forward and a bout of bargaining ensues; an average price is the equivalent of $6. A police station sits on the corner and there were police on patrol, but I saw three fist fights between boys in less than an hour, a sight I have never seen in any city except Moscow and Chicago. A more flamboyant center of vice is on a nearby street known as the Grosse Freiheit, lit by glaring neon and lined with bars and cabarets. The most famous, that in which scantily clad fat women wrestled with one another in a pool of mud, sometimes spattering the spectators, has been closed down by the city fathers. Surviving, however, is a joint where a girl does a strip-tease on horseback, as a mangy horse trots and gallops on an improvised dirt ring. The "conventional" cabarets have sights so unexampled that they make Port Said or prewar Shanghai look like Radcliffe. Stripteasing down to the bone is virtually continuous on a stage protruding into the room; onlookers hang onto the edge with their eyes popping. The bars alongside are jammed with performers between their acts and coveys of house girls, some of them youthful and attractive. In one spot I saw a Negro girl, a Jamaican, stark naked, pull out a hair or two from her most private region, and paste it with a flourish on the forehead of the stout burgher buying her drinks. In another a young woman performer jumped on a stool, sweating after her dance number, stripped off the bottom of her bikini, and smeared it over the face of a man at the nearest table, with the jovial, coy words, *"Schmutzig, nicht war?"*

This kind of thing goes on 365 nights a year. No wonder many Hamburgers seem to have a somewhat exhausted look.

Wiener Blut

Vienna, as everybody knows, is an altogether different bill of goods—a soft city, discreet, somewhat leaden, but with its own peculiar gloss and sophisticated varnish. Vienna has always meant a good deal to me if only because I spent my most formative years here in the 1930's, when I witnessed two spectacular events close up—the bombing of the Socialist houses by the artillery of the Dollfuss government in February, 1934, and the abortive Putsch made by the Nazis some months later, when Dr. Dollfuss, the Chancellor, was murdered by plotters who gained entry into the Ballhausplatz, Metternich's old palace, by a daring stratagem.

Now I prowled around for a day or two, but my preoccupation was the present, not the past. The city has not changed much in the last few years, except that the remnants of the damage done by American bombing from the air and Soviet artillery during World War II have been cleaned up. Perhaps the Viennese by and large have become more pedestrian in appearance since the war. The city has lost a good deal of inner substance if only because the Nazis killed off its Jewish citizens, who helped make it one of the most dynamic and creative capitals of the world. This was the city of Freud, Adler, Schnitzler, Mahler, and of remarkable hostesses who nurtured talent. There is less fastidiousness. Patrons of the Opera resemble audiences in East Berlin or Moscow. Women look as if they were hard put to it to keep up appearances, and men's suits seem to be made of wallpaper, a familiar Teutonic characteristic. The pace of the city is still extremely slow. It is impossible to hurry the Viennese, which is one reason why the Hapsburg Empire fell.

This left, however, a backdrop of imperial magnificence which many visitors do not seem to appreciate fully. Seeing that much of the city is down-at-the-heel, though neat, they do not accept at full value some of the authentic splendors of the past. Surely no capital in the world (except Paris, Leningrad, and the old Peking) is so lavishly, majestically laid out. The Ringstrasse, which bounds three-quarters of the Innere Stadt, is still one of the supreme boulevards of the world, with its multiple separate channels for pedestrians, horsemen in the old days, park benches, and lines of trees, as well as gently coupled tramways and streams of motorcars. I happened to arrive on a holiday; most cars, even the taxis, were decked with flowers. Everybody smiles.

The Opera, which has been reconstructed handsomely since its demolition during the war, occupies a whole city block on the Ringstrasse, and so does the gray square fortress of the Burg Theater, which is, I suppose, the most imposing theatrical structure in the world. Then, too, the Ringstrasse has the Parliament, the City Hall, the great museums, the University, the Votivkirche, and the Hofburg. This palace, which housed the old court, rises out of a backdrop of shimmering green and is, to my mind, almost as impressive as the Louvre, and has more delicacy and charm. Nearby are the Spanish Riding School, Augustinerstrasse, and the Josefsplatz, which, with its lacy neighboring streets leading toward the Opera, are unparalleled for grace, dignity, and the essence of baroque.

Nor do many visitors realize that within this antique shell Vienna is ruled by a socialist administration; this is the foremost socialist metropolis in the world in the Western sense of the word "socialism." The *Gemeinde Wien* (municipality) is, or should be, known everywhere for the multiplicity and quality of its good works—schools, clinics, kindergartens, breweries, baths, hospitals, institutes, and such immense housing projects as the Karl Marxhof, almost a mile long, the Goethehof, and the George Washingtonhof, named for an eminent and familiar revolutionary. But the impact of the municipality seems to be less direct and comprehensive than in former years. Citizens, even the most devoted Social Democrats, are lax in their attendance at *Gemeinde* festivals, and the May Day parade brings out fewer marchers year by year; the militancy of the proletariat is much diminished. One reason for this is a rise in the standard of living. At one enterprise I visited, a thousand employees out of fourteen hundred have their own automobiles. Another is the proverbial Viennese fecklessness and *Schlamperei*. Nobody cares. The old joke is still relevant: "Once it was pretty good with us and naturally it's much better now. It would be better if it got pretty good again."

The Bristol, where I stayed, is an admirably run hotel; this was the establishment occupied by the Americans after World War II. It has been refurbished and is full of pleasant small amenities.

The chair in my bathroom was cozily covered with turkish toweling, the telephone contains a radio, and you can unlock and open the front door by a clicker at your bedside. Every morning the room service waiter entered muttering "*Wunderbar . . . wunderbar . . .*" to himself, but I never found out what he was referring to—life in general, the sunshine full of ozone streaming in through the shutters, or the poppy-seed rolls and honey on his tray.

The Opera is still superb—probably the best in Europe—and few state theaters can rival the Burg for classic drama. I saw a stunning *Salome* with a youthful soprano, Anja Silja, who is much talked about in Vienna for her good looks and talent and who should go far. The theater in general is the best in Europe, after Moscow and London. *Wiener Blut* was playing at the Volksoper, and *The Merry Widow* at the Theater an den Wien; I felt that I might still be living in the age of Strauss and Lehár. But one avant-garde theater was showing *Die Seltsame Tirade des Lee Harvey Oswald,* and another held a highly modern drama based on the assassination at Sarajevo which set off World War I.

There are all manner of night clubs, including one on the Ringstrasse which has almost as many pretty girls as did Zelli's in Paris years ago. The girls of the demimonde are of a totally different genre from those in Hamburg. Even though they may be prostitutes, they act like ladies, and will accept attentions only from somebody they choose to like. But more characteristic of Vienna than the large, glittery night clubs, more attuned to its familiar semidemure sophistication, are the intimate small bars on streets off the Kärntnerstrasse, where a pianist plays continuously—any tune of any nationality anybody can think of—from early evening until the smallest hours. You sit at the bar or a table, munch hot crisp almonds, which squeak when you bite into them, sip whiskey at $2 a thimbleful, and listen to the piano player. There'll always be a piano player.

Among restaurants I went to Sacher's most, feasting on its ineffable Tafelspitz, boiled beef with horseradish sauce, one of the great classic dishes of old Europe. Sacher's is, as always, a delight

—with its complicated array of small intimate salons, crimson walls, the atmosphere of magenta, gilt molding rising from ornate marble pediments, photographs of long-dead celebrities with indecipherable autographs, and flashing silver candelabra. It is surely one of the most romantic hotels in the world, but it is down-to-earth enough to serve sound Pilsner. I liked too, as always, a crowded colorful restaurant called the Balkan Grill, where a small band of musicians wanders from table to table playing the czardas and wild Balkan chants. You start dinner with slivovitz, distilled from prunes or plums, together with cerbabcici —small spiced sausages served piping hot. A much more formal establishment on the other side of town near Stephenskirche is Am Franziskaner Platz, which makes a specialty of Backhuhn, breaded chicken, as well as a truly sublime dessert called Salzburger Nockerl, a kind of cream soufflé. Then out in the hills in the Beethoven country a few miles from the center of town the *Heurigen*—wine gardens identified by a brush of fir on the door —are full of good rough food, bounce, and rhythm, to accompany the tart new wine.

Vienna has, as almost every traveler with a tooth knows, two renowned and exquisite pastry shops, Gerstner's and Demel's. The latter makes a Sachertorte, chocolate cake with a layer of jam, closely resembling the prototype supposed to have been invented by Sacher's. The "war" between the rival adherents of these two similar confections is part of Vienna folklore. Both Gerstner's and Demel's, small, quiet, are attended by waitresses whose jobs are, I was told, virtually handed down from mother to daughter; each serves a dazzling profusion of chocolates in festive paper, fragile candies, small patisseries, and incredibly complex tiny sandwiches with delicate scrolled designs. Most of the sweet dishes carry an overload of Schlagobers, whipped cream. An artifact containing more calories than anything I have ever seen is an Obers Omelet, a cold sugared crêpe folded over half a pound of cream whipped solid as a pillow.

Vienna has nowadays something unknown in my early time, a casino. Established atop a luxurious restaurant, where the service is, alas, fantastically slow, in a former Lichtenstein palace, this seems to be the principal haunt of the local café society and show business people. Roulette chips are 50 schillings ($2) for the smallest denomination, which is expensive for Europe. Coffee houses, which continue to be the unalterable vital center of Viennese life, are everywhere. The visitor—or citizen—can sit for hour after hour over a single cup of coffee, which costs about 25 cents, while reading magazines and newspapers from all over the world, supplied free. Coffee exists in twenty or thirty different forms, from Mokka (jet black) to Weissen Ohne, which contains milk but is not topped by Schlagobers.

I had forgotten some piquant items in Viennese coffee-house lore and other customs, but they were quickly restored. In a café you normally get *two* glasses of water side by side—I don't know why—with your coffee; this drinking water is the best in Europe, coming from the Styrian hills ice-cold on even the hottest day. If you have a meal in a coffee house—or in a restaurant, for that matter—you must tell the headwaiter how many pieces of bread you have had when you are paying your bill, and these are charged for. Three different waiters must be tipped; 50 percent of the whole tip to the *Herr Ober*, who counts up the bill but does nothing else; 40 percent to the man who actually serves; 10 percent to the bus boy. This rule holds good even in the most humble cafés. Lightning mathematics is required. Incidentally, dining at a person's house, you must—by immutable tradition—leave a schilling for the cook.

Vienna has nine daily newspapers, more than any American city by a large margin, and I do not quite understand how they survive since, it seems, nobody actually buys a paper, but reads them in their rattan frames in the coffee house. I sought out other aspects of the local scene. Antique shops are abundant and full of treasure. Prices in the regular shops are comparatively low; this is not an expensive city. The Wienerwald is still entrancing; surely no other city is bounded on three sides by a hilly forest so dulcet as well as genuinely dense. Walks here along marked trails are a Sunday must. The Danube, which is seldom seen by anybody, because it flows northeast of the city beyond the Prater, is still yellowish-brown, not blue. What visitors may see near their hotels and think is the Danube is a canal. The old Musikvereinsaal still provides concerts as good as any in the world, and a new museum of modern art, charmingly situated in a park, contains Klee, Kokoschka, and a few bits of Egon Schiele.

Practically speaking, the city, which has a population of 1.6 million, has no crime. Some cops wear armbands telling what foreign language they speak, if any, as a help to tourists. There are no slums, no smog, no housing shortage, no polluted water, no rioting students, no beatniks, no fear of Communism, and no serious problems in drug addiction or juvenile delinquency. Political expression is effervescent, but most issues are settled by compromise on a mature and sophisticated basis.

The city has, it is true, minor traffic jams, but nothing like those in Rome, Paris, or New York. A Viennese friend asked me why we in New York, so technically competent and advanced, did not do what Vienna does and provide underpasses, full of shops and other amenities, for pedestrians at the most crowded intersections, as at the Opera corner of the Ring. He asked me, too, why we constantly tear our streets apart with ferocious persistence to repair gas and water mains; such a phenomenon is unknown in Vienna. I never once heard the sound of a drill piercing stone or asphalt. I suppose, too, that I should report that I never saw a miniskirt. I asked why and received the answer, "We're always a year or two behind." But not in everything.

7. S. P. Q. R.

The most remarkable thing about Rome, apart from its fabulous quality of appeal to the senses, is its continuity. This radiant cinnamon-colored city, where street signs are carved in marble, has a timelessness almost beyond measure. Here is a city which commanded the greatest empire the world has ever known, where Jews and Christians were once the same thing, and which brought the lifeblood of Greece to the world as well as contributing its own. Both London and Paris were its colonies two thousand years ago, and Rome still ranks with them as a major capital. It is one of the most venerable cities in the world, although one of the youngest capitals of a modern nation—younger than Washington, D.C.

Here we face a paradox. Rome is indeed a capital, indeed a double capital—both of the Italian state and of the Catholic Church—but it is also a very small town in spite of its population, which is around 2.6 million, by several interesting criteria. Actually, the case may be made that it is not a true capital at all, since contemporary Italy is not a thoroughly integrated nation. Rome is the seat of government, but this is a strongly sectionalist country, and other great cities like Milan, Turin, Naples, Florence, Venice, are entities that seem to have little to do with Rome, certainly little subservience to it. Local patriotism, known as *campanilismo,* is pronounced, and one community is apt to be indifferent to the rest. The local argots differ sharply. I know an Italian-speaking Englishman who, during the war, had to act as an interpreter between a Sicilian and a man from Lucca in the north. A well-known anecdote describes a man applying for a visa somewhere. He is asked if he is Italian and he replies indignantly, "Certainly not—I'm from Piedmont!"

But the Vatican, which is probably Rome's most important constituent, is a true capital. I have heard it said that the mere fact that the Vatican, an independent state, lies within Rome is what makes it impossible for Rome to be a genuine capital on its own. The Vatican is a capital—not merely of Roman Catholicism but, symbolically at least, of Christianity itself. The Italian Government of the day and the Vatican are, as we shall see, closely interlocked. I asked an experienced American news correspondent in Rome what the composition of his file was, and he replied, "Sixty-five percent Vatican, 30 percent Sophia Loren, 5 percent all the rest, including politics and government." On the other hand, even containing the Vatican as it does, Rome is not a religious city. The basic note is self-enjoyment.

Certainly this is not a city which gives easy answers. You have to dig here. There is always something beneath everything, and not merely in the sense that the most modern villa may lie directly atop the most ancient tomb. Consider the fact that Rome is both 99 percent Roman Catholic and 25 percent Communist, which means that a great many citizens are both Catholics and Communists at the same time. How is this possible? Of course one answer is that many Roman Communists, although nominally Catholic, are not very good Catholics—perhaps not very good Communists either. In Italy as a whole the Communist Party gets about eight million votes out of a voting population of some thirty million, and is the largest Communist Party in the world outside the Iron Curtain and China.

Rome is also the capital of the only democracy except Israel surviving in the Mediterranean basin, given the assumption that France under de Gaulle is no longer a truly democratic nation. What runs Rome is a combination of the Vatican and the bureaucracy. The government is a center-left coalition between the Christian Democratic Party, which represents the Vatican, and two varieties of socialists, the moderates and the Nenni group.

The Communists do not share in government, which makes politics touchy in the extreme.

Rome is distinguished not only by its continuity and sumptuous history but by the fact that it still contains tangible remains of almost every period. The past is a living reality to Romans, not something out of textbooks. Pagan Rome is still to be seen almost everywhere, as a glance at any number of antique monuments will show. There are even traces of the Rome of Romulus and Remus. Evidences of early Christian Rome are also clearly visible, to say nothing of the Rome of the Middle Ages; one need only cite the scores of monasteries which still house most of the great religious orders of the world, like the Franciscans and Dominicans; part of the city still has an atmosphere of cloistered withdrawal. The era of a later, more decadent phase of Christianity is also well represented, as is the period of baroque. Post-Reformation Rome still has its contemporary expression in such phenomena as anticlericalism on the left; the issues of the French Revolution are still being argued in the university. Finally comes the Rome of the Risorgimento, which ended the temporal power of the Popes in 1870-71. From Trajan's Forum to the hideous white memorial to Victor Emmanuel II on the Piazza Venezia, known locally as "the wedding cake," is a long gamut, but its every stage is represented by surviving stone or otherwise.

The continuity of other characteristics is similarly remarkable. A man sitting next to us in the Vatican one day had exactly the nose of the Botticelli "Ritratto Virile" in the Uffizi in Florence, a nose which I did not imagine could ever have been duplicated.

Rome is inexhaustible—another obvious quality. So far there have been no fewer than 230,000 books written about it; its churches number 800. It has a peculiar quality of universality; nobody need feel a stranger here. The whole city is a museum, one of the foremost in the world. Then, too, consider—merely in passing—its incomparable russet loveliness, its gleam and soft sparkle, the grace and excitement of almost every street leading to a fountain, a rococo stairway, a church, a cupola. Some of it seems shabby. I heard one pointed remark: "What Rome needs above all is a paint job, but this would destroy it." This city has a tactile quality unmatched by any other; put your hand on the stone of a bridge an hour or so after nightfall in early spring; it is still warm to the touch, and a quality of life rises out of it so that the contact becomes extraordinarily alive and sensual. Some streets twist so narrowly that they resemble canals in Venice; cars prod forward negotiating their eccentric turns like gondolas. Many streets have no sidewalks, and most are cobbled with stones so worn down that they have the look and almost the texture of alligator skin.

Finally—among these preliminaries—Rome is marked by a number of dazzling contrasts. I have seen a nun begging outside a shop on the Via Sistina where fabrics cost $100 a yard.

Bismarck once said, "Romans have large appetites but poor teeth." This, too, is a characteristic, because this is a city where the promise is often high and the performance nil.

Amenities Like Food and Drink

Be this as it may, Rome is one of the pleasantest cities in the world to visit or live in, even if "running" water in an apartment means dribbling water, and even if it is impossible to get the roof repaired without a bit of bribery or corruption. Servants smile as if they meant it, and a friend becomes a virtual cousin; once you enter the good graces of a Roman family, you become a member for life. Amenities are many. There's a bar as you go into the National Museum, and bookshops have chairs in which you can sit and read. In fact, almost every shop has a chair or two into which the exhausted husband may sink back while his wife looks at hats shaped like tea caddies or brilliantly original costume jewelry. The general mood is almost embarrassingly relaxed. One evening in a café at the Piazza del Popolo I saw a fat, oldish Italian sound asleep over his glass of beer—something that would be almost inconceivable in a city with as sharp edges as Paris— and the next day at Doney's I sat next to a younger man who was similarly sleeping placidly at his sidewalk table at 1:30 P.M.

Rome is, it might be added in passing, not much of a city for sidewalk cafés except on streets like the Via Veneto, frequented by tourists. It has little of the quality of Paris or other Mediterranean cities in this respect. The reasons appear to be several. Most Italians are too restless, too volatile, to care about sitting for a long interval over an *apéritif*. Second, when they go out, they go out to eat seriously rather than merely taste and sip. Again, they like to take their families with them, and many families are too big to sit at a café table comfortably. This is not to say that Romans do not like to drink. They drink like fish. Sometimes groups of men gather in the back indoor room of a café, and almost invariably look like the most sinister of conspirators.

Restaurants exist in every category all the way from George's near the Excelsior, which resembles Twenty-One in New York, to the humblest of *trattoria,* from any of the several Alfredo's, where fettucini (noodles) are the triumphant specialty, to Broadway-like joints serving non-Roman dishes like chopped liver and pastrami. Hot dogs and hamburgers are everywhere. The pizza, however, does not seem to be as well known in its native habitat as in New York. About thirty different kinds of *pasta*—spaghetti, ravioli, tagliatelli, et cetera—are available in Rome. Fruit and cheese are superb, and one of the best desserts I ever had is a chocolate confection known as *gelato tartufo* at the Tre Scalini, an establishment on the loveliest of all Roman squares, the Piazza Navona. But I felt that Italian bread was better in New York than in Rome, and for a perfect *osso bucco* I would recommend any of several Italian restaurants in Manhattan above anything in Rome. The best drink I had was colloquially known as a Kambusa—orange juice with a touch of pernod. Once, with considerable pleasure, we had a mixture of orange juice and champagne, but our host explained that this combination is not to be recommended unless the champagne is opened the day before; otherwise the concoction is too bubbly. A curiosity is that courses are served at a typical Roman restaurant with insane speed, whereas it takes forever to get the bill.

One popular bourgeois restaurant has menus translated into English. I noticed:

> Preasant with Red Wine Sauce
> Brest of Veal Rost
> 2 Quailes Devil Style
> Slice of White Fish as You Like
> Chustnuts Cake

Romans have, it would seem, a distinct sweet tooth, particularly for ice cream and chocolate, and many bars and coffee shops advertise the different kinds of chocolate available by placards hung outside. A *coppa surabaia* (a plain dish of ice cream) costs 150 lire (24 cents), and a *cornetta olimpia* (a cone) the same. A *cornetta giamaica* (also a cone) is cheaper, and a *croccante* (Eskimo Pie with nuts) is 100. For 50 lire you can get a *lemarancia*, a chocolate-coated pink stick, or a *cremino*, without the pink. A frozen ice-cream sandwich, or *zatterino*, costs 70, a *cassata* (wedge of ice cream) 180, and a *tartufo* (sundae) 80. More expensive are full-dress cakes like the *torte panna* (1,500) and the *torte cioccolato* (950).

Beatnik haunts closely resemble those in other capitals—crowded, vulgar, noisy. The most popular is the Piper. Cabarets and night clubs are on the muted side, because of Church influence; the strip-tease shows are elementary. Prostitution—despite the Church—is conspicuous all over the city, although outright bordellos have been shut down; many of the chic girls in the Via Veneto neighborhood have their own automobiles, usually convertibles, and, as in Paris, solicit trade day and night as they coast up and down the streets. There are also a considerable number of male prostitutes.

Some Moods and Qualities

Rome is, of course, built on seven hills (as are Lisbon, Amman, Kiev, Kampala), but these do not amount to much. All lie on the east side of the Tevere (Tiber), a miserably turgid stream which bisects the metropolis. The highest of the hills, the Capitoline, measures only 59 meters (193.5 feet); the lowest, the Esquiline, 31 meters (101.7 feet). More considerable are two hills which are not, for some mysterious reason, considered to be among the classic seven, but which give the visitor a good deal more impression of height—the Monte Mario and the Janiculum.

The correct official name of the municipality is Comune di Roma, and S.P.Q.R. (Senate and People of Rome) is still a symbol. Set in the campagna like a coin except on the southwestern quadrant, where a chain of industrial suburbs leads to the sea, fifteen miles away, the city has extraordinarily sharp edges. The industry in the suburbs is an inheritance from Mussolini, one of whose ideas was to push Rome outward. But in general Rome does not leak out into suburbs, like Paris or Philadelphia, but cuts off short. Immediately beyond the outer boulevards comes clear, open country where the fields bloom with red poppies in the shadow of tenements. It is as if Chicago were chipped off at the edges and planted in the middle of Kansas. The campagna stretches serenely to the Alban and Sabatini Hills, with their dusky purple glow. There is scarcely a house, shed, or hut—just open fields. I was astonished by one phenomenon—young women standing alone in the dusty grass alongside the lonely roads, alertly waiting. They are prostitutes who hail cars—particularly trucks—passing by, and offer their services to the drivers then and there.

Seldom do observers appreciate how southern a city Rome is, and how closely it lies to the Mezzogiorno (literally "midday"), the lower third of the Italian peninsula which, because of its poverty, primitiveness, and lack of education, has always—like the American South—been the problem child of the nation. The border of the Mezzogiorno is only about thirty-five miles away, which makes Rome an inevitable target for migration.

The population rises by about 80,000 a year; roughly 40 percent of this increase is the result of natural births, 60 percent from migration. The climb has been sharp recently. The population was only about 75,000 in 1870, rising to 450,000 when Mussolini came to power in 1922, and to 900,000 at the end of World War

II. Since 1950 it has more than doubled. One result is severe
overcrowding, which makes for too much pressure on the public
services and a critical shortage in facilities for education. There
are three people for every two rooms, and only one toilet for
every five citizens. Places to live are so scarce that middle-class
families find quarters almost everywhere; legend has it that there
is an apartment actually inside the arch on the Piazza del Populo.
Yet, peculiarly enough, rents are not particularly high. Most
Italians do not mind being packed together. They cling together
like Japanese. Some of the finest apartments in Rome are pent-
houses fabricated from the top floors of shockingly dilapidated
slum buildings. One somewhat scandalous phenomenon is congestion at Verano, the principal cemetery. Cremation is rare in
Italy, and embalmed corpses, for which there is no space, re-
lentlessly pile up. There are about eight thousand awaiting burial
at Verano stacked in storage vaults.[1]

There are several reasons for the steady migration into Rome
from the south. Unemployment is heavy in the Mezzogiorno, and
villagers flock northward to find economic opportunity; they have
a traditional yearning to live in towns, and they seek better education for their children. Many southern Italians are good masons,
and most become quickly absorbed into the Roman labor force
as building workers. The migration began to be accelerated by
the war, because Rome was theoretically an open city and people
flooded into it to escape bombing. (Actually, Rome *was*
bombed once by the Allies, but only once—in July, 1943.) Then,
too, the collapse of Fascism brought thousands of minor officials
and the like into the city because they could the more easily
hide here or otherwise escape vengeance from the Resistance—
Rome is a lenient city. Many migrants did not stay permanently,
but moved on to Turin and the north to get jobs in industry; this
has markedly assisted national unity. Half of Turin is southern,
and Rome itself benefited from this process by becoming more
diversified. Finally, families who lived on the outskirts of Rome in

[1] London *Observer,* July 8, 1967.

older days moved into the city to escape banditry and malaria.

Rome has no skyscrapers, one reason for this being archaeology and the nature of the soil. Deep excavation is impossible partly because of seepage from the Tiber, partly because digging might disturb antique works of art not yet unearthed. For the same reason Rome has only the most primitive of subways. When this was built, an archaeologist had to be in constant attendance, and work was stopped if it seemed that the tunneling might destroy any ancient ruins.

One day we went out to see a slum on the eastern edge of the city, near the Via Casalina. This, an ancient Roman road, was the highway along which the American Fifth Army marched to liberate Rome in June, 1944. Here miserable dwellings are built in the arches of an antique viaduct; most, about the size of chicken coops, are made of splinters of wood and plaster, with rough stones holding down roofs made of flattened petrol cans. They have electricity, but no sewage disposal; cactus grows in apertures on the roofs, and marijuana is cultivated in tiny gardens. Odd bits of boxes serve as partitions. The atmosphere is that of the rancid shantytowns on the outskirts of South American cities like Lima or Santiago, but here the area is much smaller. The squatters pay no taxes, but they cannot be dispossessed or evicted because by Roman law a man may not be forced to leave his own dwelling.

One striking fact was that the area seemed to be comparatively clean, something that cannot be said of Rome as a whole. In fact, the city was described recently by its own Commissioner of Health as the filthiest in the country. The Tiber is contaminated, which causes widespread hepatitis, and there are still about two hundred cases of diphtheria a year because of an "administrative breakdown in inoculation records." If a strike of garbage workers occurs, families even in the best districts simply toss their refuse into the Pincio, one of Rome's few parks. This is not a green city, like Brussels, Paris, or Vienna.

All traces of Mussolini have disappeared except two. His name still surmounts the proscenium in the Opera (fifth-rate in comparison to La Scala in Milan or even the Naples opera), and a square-topped obelisk still stands in what is now known as the Italian Cultural Center bearing his name with "Duce, Duce, Duce" inlaid in mosaic on the pavement and a series of stone memorials, like tombs, surrounding it to commemorate such events as the founding of the Fascist Party, the creation of *Il Pòpolo*, Mussolini's newspaper, and the battle of Vittorio Veneto. This compound is still sometimes called the "Fòro Mussolini"; it was left intact immediately after the war because it was being utilized as a rest camp for Allied troops.

✧

There are three different police forces in Rome, two of which are municipal—the ordinary cops and the helmeted force dealing with traffic. The national police are the well-known *carabinieri*, who always walk in pairs, with their swords and fancy flamboyant hats. Serious crime is comparatively scant—there are only about fifty homicides a year—but thefts and robberies number forty thousand.[2] There are no municipal courts. An offender is a charge of the nation, not the city. Habeas corpus does not exist, as is true in most Latin countries, nor does bail in most cases. An American journalist involved recently in a minor scrape spent nine months in jail before he was brought to court.

Education and public health are similarly national, not municipal, responsibilities, with certain exceptions. Figures for education are shocking. Public schools exist in Rome, but the great majority of children, like those in the most remote villages, have access only to parochial education. One afternoon we had a dozen university students to a round-table conference and question-and-answer session in our hotel. They were a mixed bag —soberly dressed, on guard at first, dark, with clever minds and inquisitive eyes. They said that only about 10 percent of the student body at the university, some fifty thousand in all, attended classes regularly, and that degrees were nothing but

[2] Again, think of the New York murder rate by comparison—653 in 1966.

scraps of paper. Only about 3 percent graduate, and a boy or girl—about one-third of students are girls—on a scholarship receives a stipend of 1,000 lire a day ($1.60) if the student lives outside Rome, half of this if he or she is a resident of the city. Textbooks are free, and meals cheap. Ten to 15 percent of the faculty—these youngsters said—were Communist; the ratio among students is about the same. An additional 40 percent are distinctly on the left. Physics, architecture, and belles-lettres are the fields which attract the left-wingers most. Only about 30 percent of the student body, although almost all are nominally Catholic, have ever been baptized. Of the boys present seven said that they preferred parliamentary democracy as a political system, one voted for Fascism, two for a "presidential republic," one for moderate socialism, one for the extreme left. I asked what they thought of divorce, which has become a rising issue in Rome recently. They were unanimously against it. I asked what writers had influenced them. The answers were, among Italians, Curzio Malaparte and Giuseppe Tomasi di Lampedusa by a large margin; nobody seemed to have much interest in such contemporary figures as Moravia, Silone, or Carlo Levi. There was one vote for Pirandello, none for Croce or Pareto. I asked what foreign writers were the most popular. Hemingway led by a long way; then came Arthur Miller and Boris Pasternak. Talking to these boys was exceptionally difficult, because I had forgotten that almost all Italians make long fiery speeches not only in answer to a question but even in asking one. After a slow start everybody talked at once.

The youth in Rome wear clothes somewhat more conservative than those in London: in Rome, as I heard it said, you can still tell a boy from a girl. Miniskirts are not microskirts, if only because most Roman girls have somewhat ungainly legs. Trouser suits, more becoming, are favored instead. There are comparatively few professed hippies; these wander the streets aimlessly, and some sleep on the Spanish Steps in the Piazza di Spagna. Sports and, of course, automobiles are the major interests, indeed fervid preoccupations, of most youths. "La Dolce Vita" as such is considered to be dated. One rising concern is drugs. Youthful Italians, following the mode, have sniffed glue, chewed morning-glory seeds, and smoked the lining of bananas. LSD has reached the scene, and its use has given birth to a new word—to become hallucinated is to be "elsdizzato." One point is that anybody can get almost any drug in a Roman pharmacy without a prescription. Some youngsters take such pills as Methedrine and Maxiton, which have Benzedrine or amphetamine bases, but the principal rage now is something called Revonal. Marijuana cigarettes cost about 1,000 lire ($1.60) each in Rome (as against 300-500 in Milan); hashish is the same price for a minuscule amount. Heroin, a true "luxury" item, is more expensive.

Servants have become somewhat scarce; their wages have risen (up to $110 a month for a good cook), and they are difficult to keep. Maids no longer want to live in, which is part of the emancipatory development now going on almost everywhere in Europe; they say that to live with a family deprives them of "psychological room." A servant of the old school is something special, and becomes a quasi member of the family; this means that she insists on being part of everything that goes on, and must have a shoulder, that of her mistress, to lean or weep upon, at the same time being a rock of fidelity and strength herself.

Rome has, incidentally, its own version of "John Smith." He is "Giovanni Rossi," and there are sixty-six of them in the telephone book. One curiosity is that telephone numbers exist in any number of digits from three to seven. The palace of the President is at the top with three; major embassies and ministries have four; Giovanni Rossi has to be content with seven.

Taxes are complicated and numerous; something known as the "family tax," based on such factors as whether or not a household owns a piano or keeps a dog, defies rational analysis; it drives citizens mad, and leads to a fantastic amount of subterfuge. The average per capita income in the city is about $1,150 a year, but this figure is slowly rising year by year. The municipality is flat broke, suffused with corruption, and heavily in debt.[3] Recently it had to borrow substantially from the social security funds in order to pay for the gasoline necessary for running the buses. Meanwhile prices continue to go up.

Economically speaking, Rome is a parasite; it lives on the bureaucracy, tourists, and the Vatican, in the sense that the Vatican brings in countless pilgrims and other visitors. It has practically no industry, even though Italy as a nation has rapidly become an important industrial country since the war. There are, however, as mentioned above, light industries on the Roman outskirts —plastics, sheet metal, fancy textiles, pharmaceuticals. Not less than one-quarter of the city's total population works for the public administration, that is, the government. Here are some sample wages and salaries per month in dollar equivalents:

Mailman	$ 130
Shoe salesman	133
Elementary schoolteacher	150
Traffic cop	152
A cardinal	650
A general in the army	747
A cabinet minister	2,000

And prices:

Rent per room	$ 9.95
Cup of coffee in a café	.40
A good FM and automatic radio	74.50
A pair of sunglasses	12.60
Babies' shoes	2.80–3.50
Kilo of apples	.25
Kilo of veal	1.30
Pair of overalls	12.50
Newspaper	.09

Rome has become known as "Hollywood on the Tiber," and indeed Italian movies have been spectacularly successful almost

[3] A recent article in *Encounter* says that the debt is so huge that "its income no longer suffices even to service the interest on it."

everywhere—such names as Loren, Mastroianni, Lollobrigida, Cardinale, Rossellini, Fellini, Visconti, De Sica, De Laurentis, Ponte, Antonioni, are familiar all over the world. Italians are passionate moviegoers: the gross box-office return last year was the equivalent of $265 million, larger than that of France and West Germany together. The pictures shown are 50 percent American, 40 percent Italian, the rest scattered. The industry is centered at Cinecittá on the outskirts of Rome, which, contrary to general thought, is not a community but a single large studio (four pictures can be made simultaneously) owned by the Italian state; it was founded by Mussolini to give opportunity to one of his sons, who wanted to go into movie-making. By American standards it is not a first-class studio. Americans began to make pictures here partly because labor was cheap, partly because so much talent was readily available. American producers and distributors put about $20 million a year into the Italian film industry, and take about $25 million; on the other side of the fence Italians take about $10 million from the United States. A new phenomenon is the "spaghetti western," movies financed in Italy but often actually filmed elsewhere—for instance, in Spain—for release to the United States. Censorship is somewhat severe in Italy, both for foreign and homemade films. Antonioni's "Blow-Up" has, as an example, never been allowed to be seen in Rome.

I went into a bookstore, and the title of a paperback Dickens caught my eye—*Canto di Natale*. I do not know Italian, and it took me an instant to seize on it that this must be *A Christmas Carol*. I looked through some other translated works. Saul Bellow's *Le Avventure di Augie March* was easy, but what about *Il Cammino Più Lungo*, by E. M. Forster? I had no trouble with *La Porte Stretta* (Gide), or *Il Capitale* (Marx), but Guglielmo Shakespeare's *Come vi Garba* kept me guessing, as did his *Sogno di Una Notte d'Estate*. Among pushovers were Walter Scott's *La Sposa di Lammermoor*, Bertrand Russell's *Perchè Non Sono Christiano*, Meredith's *I Tragici Commedianti*, *Il Circolo Pickwick* by Carlo Dickens, and *Il Molto Onorevole Sig. Pulham* by John P. Marquand. Ernest Hemingway's *Avere e non Avere* was not difficult, nor was *La Scimmia e L'Essenza* (Aldous Huxley), but I was troubled for a moment by *I Belli e Dannati* (Scott Fitzgerald) and another Dickens, *La Bottega dell' Antiquario* (*The Old Curiosity Shop*).

The number of compelling sights in Rome from the point of view of the tourist or the resident seeking culture or pleasure is, of course, enormous, from the Campidoglio Steps (Michelangelo) to the Venus of Cyrene in the National Museum, which I have thought for many years to be the most beautiful statue in the world. One need not even mention such standard, classic objects of attention as the Sistine Chapel, the Pantheon, the Colosseum, or the Forum. At this latter a dramatically effective illuminated show is put on at night. One modern sight is the capacious new international airport. So far as I could discover there are no vocal announcements of incoming or outgoing flights, which makes for a strange, silent atmosphere. Rome is a more central point than is normally assumed; Moscow is only 3½ hours away by jet, and Tel Aviv only 2 hours 50 minutes. You get a strong sense of Africa and the Middle East, and lines put in here that I never heard of—Garada (Indonesian) and THY (Turkish), as well as Air Ceylon, Sudan Airways, Kingdom of Libya, Malev (Hungarian), JAT (Jugoslav), Syrian-Arab, and Air Algeria. Flying on some of these lines must be an adventure.

Cucci, Pucci, Gucci—all these have to do with fashion. Gucci is a shop famous for its leather goods and other accessories; the other two are couturiers. Among fashion houses of the first rank are Valentino, Mila Schon, and above all Princess Galitzine, a Russian woman of spectacular good looks, verve, and talent whose husband is a Medici prince. The Roman style is, in general, less austere than the French, and most consonant with today's easy habits of living. The stress is on informality, comfort, color. Clothes are, on the whole, a good deal less expensive than in Paris, and you can buy them off the hook; ready-to-wear is in. Then, too, Roman dresses take up less room than French and are easier to pack, a considerable convenience in an era of air travel.

What mostly distinguishes Italian fashion is that it takes a middle line between comfort and emphasis on sexual temptation and appeal. In particular, Italian sports clothes and accessories—bags, scarves, shoes—are as exciting and satisfying as any in the world.

Rome is, as is inevitable, full of curiosities. A prominent commercial bank is named the Bank of the Holy Spirit, and a Holy Ghost Hospital exists. The city is said to contain 784,000 cats. A priest may not carry a woman, even his sister or cousin, on the back seat of his motorcycle. Both the BBC and United Press International happen to have their offices on Propaganda Street. The Via Appia Antica, which goes all the way across Italy to Brindisi, is no wider than a pool table. Street scenes can be glorious quite aside from bridges and temples—I saw one day a girl on a motorbike skimming near the Borghese Gardens, with thick auburn hair, the color of Rome itself, falling over her shoulders so slickly and smoothly that it looked like a cape. You can rent the Castel Sant'Angelo, the historic tomb which was Hadrian's mausoleum, for a cocktail party, and the Hotel Eden, where we stayed, has an ingenious device whereby you can click off all the lights in the room from bed. I heard that this admirable establishment had to change its name temporarily when Anthony Eden was the British Foreign Minister because of his acute unpopularity in Italy at the time.

Rome's greatest distinction is of course its quality as a museum, but a metamorphosis is in process whereby the city is being forced to face the hard facts of contemporary urban life—the evolution is from museum to metropolis. The circumstance that Rome is—almost literally—a museum still has, however, profound practical consequences, in that tourism is by far its biggest and most lucrative "industry." Nearly three million tourists came to Rome last year, and they spent some $240 million.

Finally, the Comune di Roma has a lot to worry about—or should have—in such realms as education, finance, and the bureaucracy. The reason why so few Romans ever emerge out of the mass is lack of education, and the reason why so few, even if they do manage to emerge, ever get anywhere is the throttling hold of the bureaucratic establishment. This is a city tragically devoid of opportunity for the youthful citizen.

What is a Roman?

There is no such thing as an authentic Roman if by "authentic" we mean a lineal descendant of a citizen of Imperial Rome under the Caesars when the Empire stretched halfway around the earth. Nobody can trace his or her genealogy before the eleventh century A.D., because no written records exist in Southern Europe. There are, however, some distinctly old families in Rome, not merely in the aristocracy but in the artisan class as well. I asked one lady if she was a Roman by heritage, and she answered tartly, "Only for three hundred years!" Another said, "I have been a Roman since A.D. 1400!"

Nowadays, of course, the original strain has been profoundly diluted by various admixtures. The story is that it takes three generations to make a Roman, and should take seven. The undisputed leader of the local society today is Lebanese by birth. Certainly the Romans, old and new, have special characteristics that distinguish them from most other Italians. One quality is a kind of superior passivity; another, contempt for anybody not Roman, even if he lives thirty-five miles away; another, a tendency to shrug things off, to scoff. A friend told me, "Anybody is a Roman who is both arrogant and ignorant."

A score of different communities, which interlock but still remain separate, make up the Roman amalgam: Calabrians, Sicilians, et cetera, stick together (as do Basques and Bretons in Paris). There are two different Roman manners of speech—*bocca romana,* the language of the educated, which is exceptionally pure, and *romanesco,* an argot heard mostly in Trastevere, the district on the west bank of the Tiber which houses the oldest and most firmly implanted of Roman communities, and has a picturesque separatist tradition. A restaurant on the other side did not have a table free one evening for an old customer, and the manager apologized, "Sorry, but a horde of Romans descended on us from Trastevere."

The Roman nobility, most members of which bear titles bestowed by the papacy, call themselves the "true" Romans, and are divided by a somewhat shadowy line between "white" and "black." The descendants of those who took the side of the Pope and banded together in an oath of nonparticipation when the revolutionary Italian forces, led by the House of Savoy, annexed Rome in 1870 are the blacks. Some families are mixed, with both black and white elements. Essentially the black aristocracy means the Vatican aristocracy. The heads of two great families, the Orsinis and Colonnas, have for generations had the duty and privilege of standing at the side of the Pope at official ceremonies. The blacks are by no means as prosperous as they once were; several families have become obliged to rent out part of their huge palazzos. On the whole, however, the Roman aristocrats do not retrench to anything like the degree that has become common in England, as an example. Most of the elder aristocrats, who still seek to live desperately social lives, hope to hold out to the bitter end as a superprivileged class. Their sons and daughters do, on the other hand, often go to work nowadays; there are half a dozen princesses who have established boutiques or otherwise earn an independent living.

One peculiarity is that comparatively few old families have a compelling interest in the details of their own heritage; in the blasé Roman fashion, they profess to ignore the past, even though it is this that separates them from the parvenu. Many senior aristocrats have large holdings in land, which they cling to in the hope, almost certain to be fulfilled, of rising prices.

A Word on Politics

The Mayor of Rome, when I was there, Signor Amerigo Petrucci, was harder to see than any political figure in my experience out-

side the Iron Curtain. First I had to send him a letter, then call
at the Campidoglio twice to be passed by his secretaries and press
officer, then submit a written list of questions, then send him
an autobiographical sketch of my own life and works, and finally
submit a recommendation from the Foreign Office. Once seen,
Signor Petrucci proved to be an amiable as well as modest man
—short, stocky, immaculately dressed, with glossy smooth cheeks,
black hair, and heavy spectacles. He became Mayor in March,
1964, after many years of rough-and-tumble politics in the Christian Democratic Party. He got his unusual first name not out of
any association with the United States but because it was the
name of his godfather.

To open the conversation I mentioned that I had once been
arrested in Rome, under Mussolini—the only time this has ever
happened to me—and that I had once been married (by a Fascist
official in a black shirt) a few yards from the office in the Campidoglio where we were sitting now. The Mayor received this
information soberly. He proceeded to tell us that traffic, housing,
and the public services were his major problems. The steadily
mounting population made it almost impossible to provide adequate facilities, water in particular, to everybody. The conversation was not particularly free, and I had the impression that the
Mayor was giving us figures on the local financial crisis more
favorable than the facts justified. He outlined the principal
sources of communal revenue, stating that taxes contributed 80
percent of the budget, but he avoided mentioning the enormous
deficit. Then Signor Petrucci described the division of functions
among the national government, province, and city. The former
handles education and public health in part, but the province
also has jurisdiction in these fields, including sanitation. The city
is charged with the maintenance of civilian records, statistics,
public safety, and some aspects of what was described as "public welfare." In conclusion he averred that the only possible solution to Rome's problems was industrial expansion. As of today 31
percent of the work force is employed in the building industry,

and he wants to diminish this unnatural proportion by diversification.[4]

The Municipal Council, which runs Rome in conjunction with the Mayor, is a reflection in miniature of the national government; most city problems are national, and issues are contested mostly on a national basis. There are no fewer than nine political parties —in the city as in the country—which gives rise to the saying that Italy suffers from "partitocracy"; no party ever gets a clear majority, and, as mentioned early in this chapter, government has to be by coalition. At present the municipal administration contains 26 Christian Democrats, the Church party, out of 80; the two socialist groups have 15, and the Republicans one. The opposition consists mainly of the Communists on the left (21) and the right-wing Liberals (9). The fact is striking that both the chief political parties in Rome, as in the country at large, should represent "foreign" powers—the Vatican and Kremlin. Another fact is that the government, even though democratically elected, seems to operate far from the people.

Cars, Cars, Cars

Traffic is everybody's preoccupation. Congestion in the streets of Rome is probably the worst in the world, and there appear to be three main reasons for this aside from bureaucracy, inertia, and lack of gumption. First, no main north-south or east-west arteries exist. Second, most streets are narrow, and cannot be broadened. Intersections of four or even five streets in the center of town look like starfish. Third, the vast proliferation in the number of automobiles.

Rome has one car for every four persons, and 260 new cars are "born" every day; that is, there were approximately 100,000

[4] Mayor Petrucci resigned office late in 1967 following a scandal involving the national government. Although no longer Mayor, he remained Director of the Municipal Budget, and planned to run for the Senate in the next elections. He was, however, arrested, indicted, and jailed, on the charge of "having misused funds of the national maternity and infants agency he formerly headed in ways designed to favor his own and his party's electoral interests." Robert C. Doty, in the *New York Times*, January 24, 1968.

new registrations in 1967. A car is more than a luxury to a Roman, more than a status symbol, and certainly more than a mere means of conveyance; it is part of the average citizen's inner being. To be without a car is to be like a prewar German without a uniform—undressed.

Eighty percent of all cars are Fiats, and by far the most popular model is the "500" or *cinquecènto*, which costs $900 new. Gasoline, steeply taxed, costs the equivalent of 80 cents a gallon. Most Romans drive with passion, and only yield their position vis-à-vis another driver at the last possible moment. If the drivers were not skillful, the population would be decimated, but, in actual fact, accidents are comparatively few. Parking tickets are paid for on the spot to the cop; the fees go to the municipality. Traffic jams are of a denseness, sinuousness, and complexity I have never known in another city. It took the British Ambassador two hours to get home from lunch the other day from a spot a couple of miles away. One solution, proffered by a worthy citizen, is that the authorities should wait until the entire city is locked into one gigantic jam, then cover it with asphalt and start all over again. Rome has, some visitors think, been "destroyed" by the automobile. The piazzas, among the most beautiful in Europe, are ruined, even at night, because they are "solid with tin"— parked cars. Practically nobody uses garages, which are in short supply. Somebody asked a distinguished citizen recently what inscription he wanted on his tombstone, and he replied, *"Ero un Pedone"*—"I was a Pedestrian," because he felt that not to have owned a car was his principal distinction.

Taxis are agile, swift, and comparatively easy to find. Many have round stubby fronts, and look like miniature buses. There were exactly 7 taxis in Rome in 1918; today, 3,056. Some drivers sing.

Plum in the Pudding

Vatican politics are extraordinarily abstruse, and depend to a large degree on the personality and attitudes of the incumbent

Pope. When I asked well-informed Romans, "Who runs the Vatican?" their general answer was incredulous laughter that I should be so naïve as to ask anything so unanswerable. But a kind of answer does exist—what runs the Vatican year in and year out is the *Roman* element in the Curia (Vatican government) and the College of Cardinals.

As is known to almost everybody, Vatican City is the smallest independent state in the world, covering 108 acres—one-eighth the size of Central Park in Manhattan—with a population of about 1,000. The original population (1929) was 532. The principal monument is, of course, St. Peter's, the largest church in the world, and, to my mind, one of the ugliest. There are also extra-territorial papal estates and enclaves in Italy outside Rome, like the Pope's summer palace at Castel Gondolfo. The name "Vatican" probably derives from the Latin word *"vates,"* meaning "tellers of the future"; long ago this area was the haunt of sooth-sayers. The word "pontiff" meant originally "bridge builder." Until its overthrow as a temporal power in 1870 the papal sovereignty stretched over 17,200 square miles in central Italy, and it was this huge block of papal property, stretching from sea to sea, which more than anything else impeded the unification of Italy. The Pope was a king with his own domain. (He is still a king, but the domain is more spiritual.) After 1871 the Holy Father withdrew to the Vatican, and relations between the papacy and the national government were not regularized until the Lateran Treaty of 1929 made by Mussolini.

Sixty-five countries maintain embassies or legations to the Holy See quite distinct from their missions to the government of Italy. Two of these are Communist, Cuba and Poland, and seven are non-Christian—United Arab Republic, China (Taiwan), Japan, Iran, Lebanon, Turkey, and Indonesia. The British always send a Protestant as Ambassador to the Vatican, because he necessarily represents the British monarch, a member of the Church of England. The United States maintained for a time a "personal representative" to the Pope, also a Protestant, but this has been given up. The American Ambassador to *Italy*, resident in Rome, has no formal association with the Vatican in any way.

Vatican City is the only state in the world which publishes no budget. The Pope has no fixed salary, but draws at will on the papacy funds, which derive preponderantly from Catholic sources in the United States. The papacy is, of course, inordinately rich—nobody knows how rich. By terms of the Lateran Treaty the Italian Government paid out to the Vatican $25 million as indemnity for the loss of the Papal States, and this sum has been considerably augmented since. In fact, the Vatican is by far the largest owner of capital in Italy, with resources in the billions; its investments are prodigious in both land and industry, and, a cogent fact, its wealth is constantly being increased because it pays no taxes to the Italian Government—not a lira. Nor are priests taxed, even though they are considered to be state employees. Most churches are national, not papal, property. The Vatican also derives minor revenue from such sources as the sale of its postage stamps, worth $400,000 last year, and ticket sales to the Vatican museums. The Italian Government quarrels perennially with the Vatican over taxes, because many Christian Democrats, even though they form the Church party, think that it is an outrage for the Church to be altogether tax-exempt. The socialists and Communists naturally share this view, but, after repeated crises, the present tax-free status of the Vatican has been maintained.

Squeezed into Vatican City, aside from the papal apartments—known colloquially as the Second Floor, which gives rise to such locutions as "Clear it with the Second Floor"—are some thirty streets and squares, mostly unnamed and carrying no signs, a railway station (to handle incoming freight), tennis courts, a grocery store, the plant of the *L'Osservatore Romano* (the papal newspaper), three cemeteries, a minuscule jail, carpentry and metal shops, a magnificent library, the barracks of the Swiss guards, and even a bar near the sacristy of St. Peter's.

The Vatican has its own post office, its own currency (of the same value as the lira), and a powerful radio station which broadcasts in thirty-one languages. Its official language is Latin, not Italian, and it issues its own passports and has its own "national anthem," written by Charles Gounod, the composer of *Faust*. There are four police forces—the Swiss guards, the Guardi di Nobile (a ceremonial body composed of the papal nobility), the Palatine guards, and the regular gendarmery. Sometimes riots take place on or near Vatican territory, provoked by anti-Church demonstrators or others, which may lead to intervention by the Roman police as well. Of the lot the Swiss guards are the most interesting, if only because they wear splendid uniforms designed by Michelangelo, with red, orange, and blue pantaloons, leg-of-mutton sleeves, a white ruff, and orange and blue gaiters; they are recruited in Switzerland, and must be Catholic. While serving at the Vatican they have double nationality. Passing a group of Swiss guards at a papal reception not long ago I was astonished to hear them mutter to one another in German. This seemed to sound a discordant note, but then it dawned on me that these men must have come from the German part of Switzerland.

As a rule the Vatican is assumed to be a monolithic institution, single, unitary, and dominated utterly by the doctrine of papal infallibility. Actually, it is not well organized by contemporary standards of efficiency, and its policy and tactics are often confused and even contradictory. Moreover, it has, as was nicely said recently, two faces, one for Italy, one for the rest of the world; perhaps it has a third face as well, that which has to do with Rome apart from the rest of Italy. The Pope, as Bishop of Rome, has a particularized interest here, and Rome presents its own highly special problems.

A principal issue today, although "issue" is perhaps too strong a word, is the attitude of the Vatican toward the peoples of Central and Eastern Europe, many of whom, although under Communist regimes, are solidly Catholic. The present Pope, Paul VI, has been criticized for what seems to be a somewhat lenient attitude toward these regimes, a line sometimes described as "supernational," or even "neutralist." Some Roman authorities fear that he may move toward some kind of accommodation with the Communist world in Hungary and Poland, on the grounds that the Church cannot survive in Eastern Europe otherwise; after all, he leads a "universal" church. Such a prospect horrifies Italian conservatives, who allege that it is being promoted behind their backs by left-wing French Jesuits; these have a considerable amount of influence on the contemporary Vatican, as have several somewhat radical archbishops in South America.

The Romans say jocularly that the Vatican has several favorite countries—to wit, Russia, China, all of South America, and most of the Mediterranean area. It dislikes Scandinavia, Great Britain, the Netherlands, and in particular Italy, although this is its own home ground. Italy is apt to be "neglected," because it does not have a national church of its own. Of course the Vatican itself is supposed to provide this, but if any issue arises which calls for a choice between the needs of the Church as a whole and those of Italy, the Church, being "universal," is inclined to play Italy down—so indignant Romans sometimes say.

Be that as it may, all Popes have been Italian since Clement VII in 1523. Of the College of Cardinals, which numbers 87, no fewer than 32 are Italian. In the Curia, Italian Cardinals number 21 out of 28. Vatican news and statistics are, incidentally, hard to track down and full of technical difficulties. The most abstruse reference book in the world is, I was told, the *Annuario Pontificio*, which has 1,798 pages of fine type.

The College of Cardinals is, in effect, a kind of club, and, as in most clubs, only a minority of members are active. Probably not more than eight to twelve Cardinals really count as far as actual electioneering, i.e., choice of a new "president" of the club, or Pope, is concerned. Relationships are based on who has done you a favor, who are the most dependable, who is going to be the easiest of access in a crisis, et cetera. The whole background—I do not mean to be sacrilegious—is one of IOU's, deals, and secret favors. One Cardinal is played against another. Nor do I mean to be offensive in writing that a large majority of the Cardinals are very old men indeed, some of them senile.

Relations with the Italian Government are managed with the

most consummate delicacy. The government, by terms of the Concordat, is obliged not merely to protect the Vatican but to preserve the holy character of Rome; to insult the Pope is a crime under Italian law. On its side, the influence of the Vatican on the present government comes close to being absolute. How is contact on a practical level maintained between the papacy and the Christian Democratic Party? This is a difficult question. Cardinals may occasionally meet members of the government on social occasions, but nobody whispers instructions in anybody else's ear. It is all much more subtle than that. The Church has, so to speak, a kind of mortgage on the government. Relations are established in the most shadowy way, almost as if by sign language within a family. A general of the army may call on an archbishop; a big industrialist may call on the general; and so on down the line. But out in the country contact is much more intimate. Below the bishop level the parish priests and the local secretaries of the Christian Democratic Party operate hand in glove. Priests greatly influence local politics, because they help get out the vote and can easily make or break the power and political career of an aspiring politician. It is extremely difficult for the average Italian to cut the tie with the Church. They pretend to scoff, but in their hearts they obey. What the Vatican wants, it gets.

Pope John XXIII, who immeasurably broadened the horizons of the papacy, was an unusual political operator. He took the unprecedented step of receiving Alexsei Adzhubei, Khrushchev's son-in-law, for an audience, at about the same time that he said local priests should keep their hands off politics and not seek to influence voters. In the subsequent election the Christian Democratic Party lost a million votes, largely of women, who said in effect, "Well, if the Holy Father can receive a Communist, it must be all right for me to vote Communist, as my husband has urged me to do all along." Paul VI could not, incidentally, ever have become Pope if it had not been for the good, stout, and earthy John, because it was John who made him a Cardinal. Affectionate stories and anecdotes about John are still heard on every hand in Rome; he is talked about—if only as a creature of legend—

more than Paul. It was thought by everybody that John would be merely an interim Pope (indeed, chosen at seventy-seven, he had little other expectation). In fact, this was why he was picked —to give a brief period of relaxation after the sharp rigors of Pius XII. It was assumed that he would be innocuous, but he surprised everybody by his resolute acumen, modern mind, active good works, and liberalism.

By comparison Pope Paul seems somewhat masked, a puzzle not yet resolved. He is a product, some critics say, of the "Milanese Mafia," meaning that his experience as Archbishop of Milan made him "temporal-minded." One somewhat bitter joke is that, on his election, it was believed that he had the brain of Pius XII and the heart of John, but it has turned out to be the other way around. Paul is a kind of Hamlet. He finds it difficult to make up his mind, and is an extremely shy man to boot. A physical point is that he has very strong arms, like those of a wrestler. He blushes easily. His smile is benign, but pointed. I saw him at close range when we were invited to a general audience at St. Peter's. The Holy Father was carried ceremoniously down the aisle on the familiar golden throne, and proceeded to greet delegations that had arrived from all over the world. Scenes like this are memorable, if only because their atmosphere is so unexpected. There were at least five thousand people packed into the church; they screamed, shouted, clapped, threw flowers and handkerchiefs. It was like a bull fight. After his introductory discourse, Paul walked about with a decided informality, giving two or three minutes of what seemed to be vigorous talk to a dozen or more dignitaries. His manner was relaxed, but I do not think that he is an easy man.

In a word, Paul gives the impression of being a person who is waiting—sitting out a difficult interval. His right-wing critics have gone so far as to call his Encyclical of April, 1967, the "Catholic Communist Manifesto," but this is nonsense. Perhaps the chief thing to say about him is that the bent of his mind is apostolic— which is one reason why he chose the name Paul—and his dearest wish is to extend the universality of the Church, if such a concept as universality can be extended. His acute interest in Eastern Europe derives from the time when, as a young man, he served as Nuncio to Warsaw. He was chosen Pope because he was comparatively young, able, realistic, knew how to keep his mouth shut, had few enemies, knew exactly how to get what paper to what desk, and was a pragmatist as well as a visionary.[5]

[5] But his 1968 encyclical (delivered after these pages were written) created disapproval and consternation all over the world in both Catholic and non-Catholic circles. It forbade any artificial means of birth control.

8. WARSAW—THE "WILD EAST" OF EUROPE

The principal sight of Warsaw is Warsaw. There is nothing here to compare with San Marco in Venice or the Tower of London, because the city itself is the major attraction. Warsaw was destroyed—destroyed in the literal sense like Carthage, leveled to the ground—by the German fury in World War II. Now it has been largely rebuilt—moreover, rebuilt for the most part on its own old pattern, a simulacrum of its former self—by the romantic vigor, patriotism, and *esprit* of its own faithful citizenry. This is a phoenix city, risen from the ashes, born anew.

There were three principal phases of destruction made by the Nazis between 1939 and 1945.

First came the initial attack on Warsaw, which, as everybody over middle age will remember, was the opening act of World War II, on September 1, 1939. The Wehrmacht struck with every ounce of force it had. Warsaw was an open city, altogether unarmed, but it held out for two furious weeks of siege until it was forced to surrender. About 10 percent of the city was destroyed outright by the combined German bombardment from the air and on the ground, and fifty thousand Polish citizens were killed. Hitler was so enraged at the quality and intensity of the Polish resistance that he determined to give the capital a real lesson. Sections of the city were laid waste systematically one by one, while German experts went through the ruins of handsome buildings determining which historical records and works of art that had survived should be stolen and transported to the Reich, and which should be destroyed. The Germans planned to build a "Fortress Warsaw" across the Vistula, where 200,000 Poles—no more—would be permitted to survive as slave workers. Warsaw had at the outbreak of the war 1,300,000 people, of whom some 500,000 were Jews. The Hitler plan to permit 200,000 Poles to be spared was soon modified, and a policy of total annihilation adopted instead. The monstrous execution factories were set up at Auschwitz, Treblinka, and elsewhere, and a new word—genocide—entered the world's vocabulary. More than three million Poles, mostly Jews, were murdered before the holocaust was over.

After more than three years of horror and infamy unparalleled in the modern world the surviving Warsaw Jews, perhaps fifty thousand in number, revolted. They decided to die fighting rather than be killed off without resistance. Thus came the second act in this miserable record—the Ghetto Uprising of April–May, 1943. The story of this is too well known to repeat here. After savage fighting street by street, house by house, the Nazis managed to put the uprising down, and most of the Jews who survived were butchered. The ghetto (in the center of the city) was pulverized into rubble. Thousands of non-Jewish Poles were rounded up and killed as well.

Act Three began on August 1, 1944—the Warsaw Insurrection. Survivors of the ghetto massacre and others were made desperate by hatred of the Nazis and starvation. The bread ration for citizens was two kilos per month; the average diet was 180 calories a day. Thus the "final solution" was being remorselessly put into effect. So, once more, the surviving Jews, many of whom were living in holes in the ground like animals, and who communicated with one another via the sewers, rose to revolt. This lasted for sixty-three heroic days. It was during this episode that Joseph Stalin, with crack divisions of the Red Army comfortably stationed directly across the Vistula—which could have cleared out the Nazis in a day or two—refused to budge, mostly because Stalin did not want a strong Poland to survive after the war. Now the Reichswehr, having at last crushed the insurrection, proceeded in earnest to finish Warsaw off. Hitler decided to erase it from the skin of the earth; his orders were: "Level Warsaw to the ground," and this is almost literally what happened. The Nazi commandant on the spot, reporting to Berlin, announced that never again would a Pole, Jew or non-Jew, live in the city. Some 250,000 were killed on the spot, and 200,000 more were dragged off to the asphyxiation furnaces; the Germans then proceeded to destroy the city building by building, stone by stone. Homes were set on fire while still full of people; citizens popped out of windows, living torches, to crash to their deaths. That all this was carefully planned is proved by eleven thousand pages of documentation captured from the Germans after 1945. Deliberate experiments were made to see whether a house burned more easily if its windows were open or shut. How to blow up a sewer efficiently became a serious subject of study. The Vistula bridges were destroyed, and the very trees uprooted in the parks; the ruins were mined.[1] A detail almost minor—compared to some—is that twenty *million* cubic meters of debris covered the city. One could not tell the site of one street from the next. When, a few months later, the war ended at last and the city was liberated, exactly *fifteen* surviving human beings were found.[2]

Other facts and figures about wartime Poland as a whole, as well as Warsaw, paralyze the imagination. Some 38 percent of the country's national wealth was lost (as against less than 2 percent in France, as a comparison). More than 80 percent of all industrial enterprises—factories and so on—were demolished. Most archives as well as contemporary legal documents were wantonly destroyed in the capital and elsewhere. Warsaw's Public Library, the National Museum, and the principal schools, theaters, and hospitals, gas lines, telephone cables, the water supply, the electricity system, were all obliterated. The fact that a small percentage of Warsaw was not razed was due mostly to the fact

[1] Tom A. Cullen in the New York *World-Telegram*, November 27, 1966.
[2] *Poland*, by Eva Fournier, in the admirable Vista series, Viking Press, New York.

that the German occupiers did, after all, have to live and work somewhere themselves. Otherwise—nothingness.

Resurrection

During most of the war refugee Poles in London and elsewhere —even some heroic citizens who managed to stick it out in Poland itself—worked busily on plans for the future reconstruction of the city. These indomitable realists—in a country commonly thought "romantic"—never lost hope. Immediately after liberation, however, when there was no gas, no electricity, no coal, no transportation or public services of any kind, and forty thousand corpses to be buried in ice-packed snow, a handful of leaders thought that reconstruction would be impossible, and that it would be best to set up a new capital at Lódz or even Cracow. (But Warsaw has always had a certain jealousy of Cracow, the brilliant and lovely principal city in southern Poland.) The decision was finally taken not to choose a new capital but to rebuild Warsaw itself from scratch, no matter how difficult this undertaking might prove to be. One irony is that this work, which began in March, 1945, two months after liberation, had to begin with *more* destruction, in order to remove buildings which, though still standing, had been so badly damaged that they were beyond repair and unsafe. Then something remarkable happened—about 300,000 Warsaw citizens who had fled the stricken city and who had survived somehow in the countryside *returned* spontaneously to their gutted capital. The next overriding decision was whether to rebuild the city in the pattern it had had before or to lay out a new metropolis under modern planning. This was discussed in a vivid eight-hour session under the leadership of the President of the Republic at the time, who happened to be an architect, and it was decided to rebuild the metropolis as it had been.

Thus began the "Miracle of Warsaw."

✼

Poland, let us remember, has what has been called a "crucifixion complex." Three times before it had endured the unique and terrible experience of concrete geographical partition, but in 1918 it rose from the dead after 123 years of extinction; this, the present (1945), crisis was, its proud and tough citizens said, merely the fourth round. But even though they made themselves sound blasé, no Pole who took part in the reconstruction will ever forget it, just as no Pole will ever forget the destruction. Not a particle of dust was touched, as one writer put it, that did not contain the "relics of a martyr." Scarcely a man, woman, or child lives in Warsaw today who did not lose a relative or friend. The President of the Republic today lost his toes, the result of torture by the Germans. Every restored building contains a large conspicuous photograph of the ruin out of which it rose, so that future generations will always remember. There was a kind of wild sporting element in the whole adventure. A good brick unearthed and assigned to its proper building was an object of adulation. One curious point is that the builders, the restorers, were cardinally helped by the paintings of an eighteenth-century Italian artist, Bernardo Belotto, a nephew of the master Canaletto, famous for his scenes of Venice.[3] Belotto, after long residence in Germany, spent ten years in Warsaw (1767-77), painting assiduously—palaces, squares, churches, streets. His work was almost as detailed and precise as an architectural drawing, and, miraculously, most of it had been hidden in the countryside and escaped destruction. Thus the Belotto paintings and drawings were available as a basis for the restoration of old Warsaw when photographs did not exist, with the result that much of the city today has the same eighteenth-century look it had before the war. The Communist authorities encouraged this because they wanted to appeal to the historic sense of their citizens and retain a continuity with the past. Similarly, in Moscow the contemporary regime, even though it overthrew the old order, scrupulously pro-

[3] Sometimes Belotto called himself Canaletto after his uncle, and is often referred to in Poland by that name, which makes for confusion.

tects the art objects in the Kremlin and encourages its people to visit and admire them.

Brick by brick the restoration of Warsaw got under way. It was still going on fervently when I visited the city in 1948, three years after the war, and I saw aspects of the "Miracle" which I shall never forget. Perhaps I may be permitted to quote from an account I wrote at the time, just twenty years ago:

> The destruction was so great that I could not find my way to the simplest objectives. Almost all the landmarks I remembered, like the lovely old Bruehl Palace which housed the Foreign Office, have completely disappeared. The Royal Castle, St. John's Cathedral, the assemblage of graceful buildings near the National Theater, restaurants like Fuggers, the pretty old round Church of Alexander, the Poniatowski monument, have been wiped off all but flat. For acre after acre the city resembles . . . a gutted moon.
>
> In Berlin, if you stand near Brandenburg Gate, you can at least see the outline of what buildings once were. There, you say to yourself, are the remains of the Hotel Adlon, there is the skeleton of a house I dined in once, there is what was the French Embassy. But in Warsaw it is impossible over large areas to identify any buildings at all, or even to see where street intersections were, because the ruin is total, the devastation is complete. Almost every vista looks like a jumble of enormous broken teeth. . . .
>
> Every Pole I met was almost violent with hope. "See that?" A cabinet minister pointed to something that looked like a smashed gully. "In twenty years that will be our Champs Elysées."
>
> The worst area is still the Ghetto, which in literal fact is a heap of rubble, nothing more, nothing less. It looks like a huge empty rocky lot. We prodded through it slowly, scampering up and down hills of crushed debris. A few straggly dandelions and cabbages, bearing pathetically valiant flowers, grow on what was once the busiest section of the city, and a clump of dusty bushes has spurted out over the area where the biggest synagogue in Europe once stood. Grass grows again. Human beings do not. And never let it be forgotten that of the 3,500,000 Polish Jews who lived in this country before the war, more than three solid *million* were butchered by the Germans. . . .
>
> In other parts of the city . . . we watched the work of re-

building. Particularly is this impressive in the Old City, which is almost as complete a ruin as the Ghetto. A patch of ravaged brick is all that remains of the Angelski hotel where Napoleon stayed. The old bricks are used in the new structures, which gives a crazy patchwork effect. Hundreds of houses are only half rebuilt; as soon as a single room is habitable, people move in. I never saw anything more striking than the way a few pieces of timber shore up a shattered heap of stone or brick, so that a kind of perchlike room or nest is made available to a family, high over crumbling ruins. One end of a small building may be a pile of dust; at the other you will see curtains in the windows.

Much of this furious reconstruction is done by voluntary labor; most, moreover, is done by the human hand. Even cabinet ministers go out and work on Sunday. In all Warsaw there are not more than two or three concrete mixers and three or four electric hoists; in all Warsaw, not one bulldozer! A gang of men climb up a wall, fix an iron hook on the end of a rope to the topmost bricks, climb down again, and pull. Presto!—the wall crashes. Then the same distorted bricks go into what is going up. The effect is almost that of double exposure in a film.

Also:

One morning we went to a bank, which was housed over a cave in half a broken building. The clerks and tellers, mostly women, were having their second breakfast, and leisurely they dealt with our checks. We chanced to turn around; there behind us on the balcony was a uniformed guard with a tommy gun. He held it ready to use, swinging it slowly to traverse the bare high-ceilinged room from one end to the other. The reason dawned on us sharply. All the vaults, iron meshing, and strong boxes of the Polish banks were of course destroyed, and so today currency is simply piled up on open wooden tables. The guard was ready for instant action. Our checks were cashed with courtesy, deliberation, and only a minor amount of red tape. But it was a far cry from the Guaranty Trust on Rockefeller Plaza.[4]

Now all this is so much part of the past that the record is all but unbelievable, although a few ruins remain. Warsaw today is a city with scarcely a trace of the old havoc and horror, a rebuilt city, neat as buttons.

[4] *Behind the Curtain,* pp. 249-256, a book I wrote about Eastern Europe in 1948 (Harper & Brothers, New York).

A Glimpse of the Polish Capital Today

Warsaw has a population of about 1,130,000 within the city limits at the moment—a bit less than in 1939. Counting the whole
metropolitan area the figure is 1,900,000. The city has what is
called a "deglomeration" program—an attempt to move industry
and its workers outside, so that the crush within Warsaw itself
will not be so pressing. Nobody is allowed to move into Warsaw
unless he or she has adequate housing, and this is difficult to
acquire. The municipality is building 35,000 new housing units a
year, but needs 50,000. The authorities are working now on a
plan for 1985, when the population of the metropolitan area will,
it is calculated, reach 2,350,000.

The Mayor is largely a figurehead; he goes through the formality of election, but is in fact appointed by the party. He was
recently discharged from his post because his daughter, it appears, had taken part in a student demonstration. We talked at
length with Zenon Nowakowski, the director of the Warsaw
Town Planning Commission, and his principal associate, Stanislaw M. Jankowski. This latter, a remarkable man whose activi-
ties have been international, is fifty-six, tall, with a splendid
Roman nose, pale luminous blue eyes, and well-kept strong
hands. Mr. Jankowski is a professional town builder; he recently
spent two years in Baghdad planning cities in Iraq, having been
commissioned to this task by the Iraq Government even though
he is a citizen of a Communist country; subsequently he was
chosen by the UN to make a master plan for the reconstruction
of Skopjle, Yugoslavia, after the recent earthquake there. Mr.
Jankowski happened to be in England studying at the Liverpool
School of Architecture when the war broke out in 1939; he spent
several years in Scotland with the exiled Polish armed forces
there, and then was parachuted into Warsaw in 1942 to be a
leader of the underground—a dangerous undertaking—and rose to
be the aide-de-camp of the Polish commander in chief during the
uprising. He returned to England in 1946, when the war was over,

but then came back to give his life and works to Warsaw—a typical enough Polish story.

Even though Warsaw was built anew only twenty years ago and even though it is a capital in the Communist domain, its principal problems resemble closely those of Western cities—to keep from becoming too big and to house its citizens adequately. The needs of industry and municipal services, the reorganization of traffic and amelioration in housing will, Mr. Jankowski thinks, continue to be the city's chief preoccupations for a good many years. Population growth must be limited, because overcrowding makes for discomfort and expense. He wants "symbolic space" for children, new shopping centers, and more cooperatives. Rent is restricted to 7 percent of the income of the head of the family, no matter what kind of dwelling is involved. This is the usual figure for all the satellite states. Central heating in the new apartments is piped in free.

Housing is almost always the principal headache in the countries of the Communist empire, as we shall see in Moscow in the chapter following. In Poland I was told (by foreigners) that the overriding reason for the continuing acute shortage in living space is not so much that proper housing is considered a "luxury," like a fur coat, but because the economy as a whole does not work well. "Housing is their Negro problem," is one remark I heard from an American. On the average a Warsaw citizen gets 7-9 square meters of living space, not much, and there are 1.37 persons to a room according to Polish statistics. No family may have more than 110 square meters. Polish statistics are, incidentally, quite reliable, something that cannot be said of other Iron Curtain countries. Cooperatives, in the Western sense of the word, are encouraged, and can be purchased; a three-room flat with a kitchen costs about 210,000 zlotys ($5,250 at the theoretical par rate of 40 to the dollar). The down payment is about 20-25 percent of the total price; the rest is paid off in "rental" installments, say 500 zlotys ($12.50) per month.[5] Of course, this

[5] David Halberstam in the *New York Times,* September 1, 1965.

is not "Communism." But not much in Poland is except the nature of the government and certain basic tenets under which the regime operates. Agriculture was decollectivized long ago and 86 percent of the total farm area is now in private hands.

Here in Warsaw we are, however, undeniably inside the Communist circle of states—the first Iron Curtain country we have visited in this book. One effect of this is drabness. Physically, most of the city, lying flat on the great Polish plain, is dreary beyond description—poor as well. The intense poverty of almost everybody was what struck me most at first sight. Old apartment buildings have jagged, broken steps, with courtyards dingy and barren to the uttermost. Another characteristic to be seized at once is that even though Poland plays hand in glove with its Communist partners on most issues—as was proved well enough by its participation in the ugly, sordid assault on Czechoslovakia in August, 1968—it paradoxically considers itself to be a Western, Catholic state as well. The population is at least 95 percent Roman Catholic now that Jewry has been largely destroyed.

What most distinguishes Warsaw and other Polish cities is the quality of the people. Poles are, as everybody knows, hedonistic, volatile, with spark, sting, and animation, and they love to talk; their capacity for conversation, most of it cheerful, some of it wildly controversial, is seemingly without bounds. Moreover, they have, in the sheerest sense, character as well as charm. Not all Poles are Artur Rubinstein, who is probably the single most delightful person I have ever met, but some run him close. There is a wild streak in most Varsovians, and they have been nicely called "God's madmen."[6]

Polish history goes way back. There are deep and tenacious tangled roots here. The Poles were converted to Christianity as long ago as A.D. 966; their first king in a long line of picturesque monarchs was crowned in 1025, forty-one years before the Norman Conquest; the university in Cracow was established 250 years before Harvard. Warsaw itself is about seven hundred years

[6] Fournier, *op. cit.*, p. 116.

old, having been founded in the thirteenth century. A small legend attends this event. Warszawa, the correct name of the city in Polish, has a mermaid as its symbol; she is supposed to have risen shimmering out of the Vistula to point out where the first settlement should be built. A fisherman caught her, but she gained her freedom, promising as the price of her release that she would remain on the coat of arms of the city and protect it forevermore. The city did not become the nation's capital until 1596, when King Sigismund III Vasa (the royal nomenclature is peculiar) moved his official residence to Warsaw from Cracow, ostensibly because the palace on Wawel Hill in Cracow had burned down, in reality because the local citizenry threw him out. The Sejm—Parliament—was already sitting in Warsaw. Officially the city did not become the capital until centuries later —under the Communist regime in 1952, when the constitution of the Polish People's Republic was adopted. Incidentally, every student of history knows that the Sejm, in its old days, was characterized by the *liberum veto; any* single deputy (out of a thousand or so) could veto legislation. This was a tribute to Polish individualism, but of course it wrecked the country. Another factor making for isolation and deliquescence was the extreme difficulty of the Polish language. Then one must mention again Polish romanticism and the exaggerations it encouraged. In Warsaw in August, 1939, just before the war, I met youthful Polish officers who boasted about how they would take Berlin in a week or so—with cavalry!

✿

Travel in Poland is controlled by Orbis, the State Travel Office, which closely resembles the Intourist organization in the U.S.S.R. The visitor buys vouchers in New York, London, or elsewhere abroad for hard currency, and these provide for hotel accommodation, meals, et cetera, within the country. We stayed at the Europejski, which I have known for thirty years or more. It has been rebuilt since the ravages of the war, and is less picturesque, less stately, than before. Equally comfortable is the Bristol around the corner, and this, I think, has better food although the menu is identical. A third hotel in the superior category, also in the middle of town, is the Grand. The Europejski has an exterior sign near the entrance, "HOTEL KATEGORII LUX," but its atmosphere is utilitarian nowadays rather than luxurious. Dominating the ground floor is a series of cafés, which are crowded from noon to nightfall with animated Poles bent head to head, talking, talking, and sipping coffee. It is difficult to find a table. Dozens of remarkably smart-looking prostitutes are usually to be seen; they are encouraged, rather than not, to sit here, because they perform the useful economic service of drawing in foreign currency if they can entice attention from a tourist. Until 10 P.M., a man and woman can, if they seek companionship, register perfectly openly for a hotel room under their two separate names. Incidentally, it is wise for a visitor to learn that △ means a men's powder room, ○ a lady's. No other signs are used.

There is, of course, as in all Communist countries, a "service" bureau in the lobby, which is a kind of substitute for the concierge, and two elevators, one of which never seems to be working; this is hung with a placard, *"Nieczynna,"* "Closed." A non-Communist touch is a bellboy (bellboys are unknown in Moscow, as an example) who silently stalks through the cafés holding aloft a blackboard with a name chalked on it—the local method of paging. The restaurants have feeble, old-fashioned jazz, and the vitrines in the lobbies show wigs for women, children's dolls, and Scotch whiskey at $4 a bottle.

The towels in the rooms are of miserable material but are embroidered; wastebaskets are covered and operate with a foot pedal; there are splendid rolls for breakfast; the toilet paper is elephant-colored, and the beds sublime. A negative point is that there is no "O" for "operator" on the telephone, and it is difficult to communicate with the Orbis people downstairs. Writing paper does not quite fit the envelope—as in Rio de Janeiro. The best thing is the quality of the old servants, most of whom are friendly and experienced stout women of the *ancien régime*.

✿

In older days Warsaw was celebrated for the quality of its cuisine; this was "the Paris of the North." Glorious hams stuffed with pistachio nuts were a specialty; so were various types of mushroom, including a jet-black variety as big as a platter, which was served grilled like a steak. These we never saw on this trip; perhaps they were out of season, perhaps they have disappeared. Specialties of the past which have survived are *chłodnik*, a delicious cold soup made of cream, nuggets of chicken, beets, and cucumbers; stuffed pike and tripe in several forms; wild boar in cream, also venison in season; and *bigos*, a kind of stew of beef, other meats, and cabbage. We ate mostly a Russian dish, Côtelette Kievsky, or Chicken à la Kiev—which, as every gourmet knows, is the boned breast of a chicken, breaded and filled with melted butter. The big hotels have menus printed in French, but going through a Warsaw menu in Polish without help can be a dizzying experience. A few words, however, retain a clue to their identity. It is not too difficult to guess *sznycel, konserwy, sos tatarski, sałata, makaron, szpinak, suflet,* and *ryz*, to say nothing of *omlet* and *szparagi*. But what about *nalesniki z borowkami* (pancakes with whortleberry jam) or *krem z. piecz. diabl.* (a heavy soup with croutons)?

A variety of light beer called Zywiec is not too bad (the bottle is marked "full" in English), and Hungarian and Rumanian wines are available. A wine bar called Fukier is one of the sights of the town, particularly in the morning when men going to work need an eye-opener. Vodka exists in any number of forms; *winiak* is a spirit matured in a brandy cask, and *jarzebiak*, like the more familiar zubrowka, but brown in color, is flavored with a reed. The best cigarettes, at 18 zlotys (45 cents) a pack, are made in the American style, and are called Carmen. As to vodka, it is quite expensive; half a liter costs a laborer's daily pay. A good deal of heavy drinking goes on, but it is comparatively rare to see anybody drunk in the streets, something not true in several American cities. Bottles of vodka or other spirits are not sold in rural areas on Saturday, Sunday, or payday, in order to discourage drinking. Citizens are, as a rule, careful not to take a second drink before driving, because, in the event of an accident, the penalty for drunken driving is exceptionally severe.

The liveliest and most amusing night spot we saw served no spirits at all, only wine. This was a joint in a cellar in the Old City, where a singer, Wanda Warska, belted out songs composed mostly by her husband, Andrzej Kurylewicz. They operate the place together. Most of the songs are newly composed serious modern jazz. One of the three musicians on the dais looked like Truman Capote, one like Jesus Christ, and one like a disemboweled Wall Street broker wearing steel rimless spectacles. This place was as authentically original and exciting as night clubs in the old Montmartre; the atmosphere combined freedom, sophistication, and modernity. I have never seen anything like it in any other Communist country. It would not be allowed to exist ten minutes in Moscow. Jammed side by side at rude tables were foreign diplomats, Varsovian officials whom I had met in their prim offices earlier that day, mini-skirted girls, boys with long hair, poets, and musicians. Prices were cheap, and laughter long.

Two tangled subjects in Warsaw are money and surveillance. There are two different legal exchange rates for tourists—24 zlotys to the dollar up to $20, then 40 zlotys for amounts higher. Obviously this is a device to encourage tourists to spend more dollars. The visitor is put in a ticklish position if, toward the end of his stay, he must calculate which will serve his interests better, say $15 at 24 or $20 at 40. Any zlotys left over are useless, since they may not be taken out of the country. In addition, there is a green-market rate, 72 zlotys to the dollar, a "modified green-market rate" (80), and the "normal" black market, which is 100 to the dollar. But an American bill in a large denomination will bring 120-130. The green rates are for funds remitted to Poland by the six million citizens of Polish extraction living in the United

States, many of whom regularly send money back to relatives in the old country. In addition, there are 3,100 residents of Poland —some 500 of them American citizens—who receive U.S. Social Security or other federal benefits.

Warsaw wages and prices are comparatively low. One little joke is that every Pole earns 2,000 zlotys ($50) a month, and spends 3,000 ($75). A good cook gets about 1,500 a month ($37.50). A ticket to the theater costs 20 (50 cents), to a movie half that. Taxi fare from one end of the city to another will rarely exceed 25 zlotys. Wages are low, yes, but rents are, as we have seen, minimal, and "visible" unemployment is unknown. Education and the social services, as well as medical care, old age pensions, and the like, are free.

In regard to surveillance, I felt more evidence of this than in Moscow. We were told later that our room in the Europejski —No. 261—was apt to be assigned to Western visitors who were thought to be "sensitive," and that it was presumably bugged thoroughly from floor to ceiling. Once a suitcase of mine was opened—the evidence was indisputable, in that the lock had been clumsily broken. Once we drove out with some friends, including a Pole, to a private-sector restaurant in the country about twenty miles from Warsaw; halfway to our destination a black Mercedes, clearly a police car, wheeled into line behind us from a side road, clung to us the rest of the way, and followed us back. An air-mail letter to London takes about ten days to get there; obviously mail is opened and inspected. My wife and I had eight introductions of a serious nature to Warsaw citizens; we succeeded in seeing four of these, so our score was half and half. Two told us over the phone that they were "ill"; from two we never had a reply. People are frightened by and large of being seen with foreigners, although some are "sanitized," i.e., cleared for contact with visitors from outside. Yet occasional bold Poles thumb their noses at the government and make pointed jokes about it. I should add that we happened to arrive in Warsaw in a tense period, just after an epidemic of student demonstrations and during a political purge, to be alluded to later; citizens were much more on guard than was usual. An American friend told us that it had been impossible for him to arrange a dinner party for us to meet some Polish intellectuals and political figures at the time we were there, whereas six weeks before it would have been a routine matter. But, even so, the atmosphere in Warsaw is still much freer *in general* than that in Moscow.

It is virtually impossible to gauge the amount of political dissent. The underground seems to be universal. One journalist told us, "It isn't easy to measure anybody's temperature, because there are not thermometers enough." The best-informed estimate I heard was that, in a free vote, not more than 6 percent of the people would favor the regime. On the other hand, scarcely anybody wants to go back to prewar Poland with its feudal miseries. What most citizens would like, it would seem, is a liberal socialist society like that in Sweden.

Municipal Problems and Such

In a talk with the Deputy Mayor I mentioned that New York City had had more than six hundred murders in a recent year, and asked him what the figure was for Warsaw. He looked horrified and rushed out of the room to consult one of his statisticians. There were no murders at all in Warsaw last year—not one.

The crime problem is, indeed, minor. Burglary and thievery are uncommon, if only because few households are rich enough to make robbery worthwhile. Last year, the total number of crimes of all kinds (in a city of more than a million people) was only 2,879, a drop of 30 percent from the year before. There are, however, a few *huligani* (youthful hooligans), who roam the streets or lurk in the doorways of apartment buildings; the favorite weapon for opening an apartment door, if the family is away, is a bicycle spoke. There are two types of police, the secret (corresponding to the KGB in Moscow) and the ordinary uniformed cops, who are known as the MO, which stands for "Citizens' Militia." Pickpocketing is a fairly common offense, and the theft of automobiles is increasing. Political corruption on the municipal level, such as distinguishes a good many American cities, is unknown.

Traffic is not a problem yet, if only because there are only 230,000 privately owned cars in the whole country. The favorite small automobile, produced locally, is the Warszawa. Commuting has, however, become a preoccupation because of new housing developments on the periphery of the capital. Seventy-five percent of commuters travel by train, 20 percent by bus, 5 percent in private cars. In all, there are about 140,000 commuters a day. The buses are clean and cheap, and some run all night; the fare for any distance is the equivalent of two cents, less for students. One minor problem has to do with jaywalking; important intersections on the streets are marked by white zigzags, on which a pedestrian is supposed to be safe, but few drivers pay much attention to these, and they are nicknamed "killer zones." Penalties for accidents are severe in the extreme; a farmer recently got six months in jail for letting his car bump into a horse.

The basis of public finance is, of course, that the government itself owns most of the shops and automatically receives their profits, known as the "turnover tax." Industries pay a proportion of their profit to the government; the percentages vary and are determined by the National Planning Board, which draws up all the budgets in the country—national, provincial, municipal. There is no property tax or *general* income tax, but professional men, the co-ops, and small merchants may be assessed a fee. In the case of a successful physician—private practice is permitted in the professions—this may amount to as much as ten thousand zlotys, or $250, a month. The biggest item in public expenditure is for education; what are described as "social and cultural activities," including education, take 69 percent of the Warsaw budget. Most of the new industries in the Warsaw neighborhood are highly specialized—electronics in particular, then automobiles, machine tools, pharmaceuticals, and radio-TV. These need sophisticated labor, which is difficult to find. Most factories are carefully situated north or south of the city, since the prevailing winds are westerly; this serves to lessen smog.

Education was shockingly neglected by Polish governments in the prewar period, as was practically everything else pertaining to the modern world. Noblemen lolled in fantastically sumptuous and ornate castles; the peasants were treated like pigs and all but starved. The country was still 25 percent illiterate in 1938; today illiteracy is virtually unknown except among a few very old people. We spent a morning one day at the Basic Metallurgical School, a technical institute for the training of skilled workers. Boys and girls are brought in at the ages of fourteen or fifteen, after primary school, if they show promise; there are 96 teachers, 1,180 students; tuition is free. The director, Mr. Szkoly, took us around, an enlightening experience. Boys and girls who go here (one of 180 technical schools in Warsaw) are almost automatically assured of a good job, paying up to 3,000 zlotys a month, on graduation, which is more than some of the teachers get, and the waiting list to get in is long. One thing that impressed me was the attention given to the amenities, though some of these are crude; a doctor, a dentist, and several nurses are in attendance, and the building, near the Vistula, is flanked by carefully tended —but modest—grass *plaisances*. Every student has to take off his or her shoes and put on slippers while at school; this derives from an old Polish characteristic—marked attention to the state of *floors;* similarly, floors are protected in museums.

In one room my curiosity was aroused by a peculiar type of tool being manufactured; it had long metal hands—a device for picking up radioactive material. These implements are exported mostly to the Soviet Union. I remembered suddenly the mechanical man which, in former days, was a prize exhibit in the Industrial Exhibition in Moscow; it could perform the most sensitive and difficult tasks, like pouring liquid from a small beaker into a test tube, without any human assistance. It was interesting to be informed that these mechanical men, with their metal hands, were first produced here in Warsaw at this very school.

Roughly 48 percent of the population of Warsaw is under twenty-five, which makes for some vivid social patterns. The youngsters don't have much ideology and are strongly influenced by Western modes. Men like to be well, even fancily, dressed; this is a city conspicuous for matching ties, handkerchiefs, and socks, even if these are of poor quality. Shoes seem to be somewhat flimsy on the whole. I saw only one man with his skull shaved bald in the Russian manner, and only one set of solid steel (or perhaps silver) teeth. Women play a large role in the economy and the professions; most dentists are women, as in the U.S.S.R. People are convivial in the extreme, and parties take place all the time (in strict contrast to Moscow), particularly during carnival, which lasts for several months and when a large shortage of sleep piles up. Remnants of the aristocracy still survive; they have not been wiped out, as in the other satellite countries, because the government does not think that the former feudal princes are important enough. Several men and women of the Potocki family, one branch of which maintained the greatest house in Poland before the war, are still members of society, and a Radziwill (Princess Izabella, whose husband is also a Radziwill, works in a publishing house. The husband, Prince Edmund, is the elder brother of Prince Stanilas Radziwill, who married Lee Bouvier, the sister of Jacqueline Kennedy. He has a job in an automobile insurance agency. All told there are still seventeen Radziwills in the country. One Potocki (Count Karol), whose large and glamorous racing stable was once celebrated, has just one horse left.[7]

Street Scenes and a Few Sights

Streets are spacious for the most part, and the city, although drab and grayish, is full of flowers—the parks are flecked with rough patches of dandelion. Warsaw loves flowers; the first shop to be restored after the holocaust in 1945 was a florist's. The main street is called Nowy Swiat (New World); this was built by an Italian, Antonio Corazzi, in the best nineteenth-century style, and a prolongation of this holds today the headquarters of the Communist

[7] Gloria Emerson in the *New York Times,* January 15, 1967.

Party, which, friends told me, is known ominously as just "The Building." Many streets are still traversed by horse-drawn carts with rubber tires; the rear wheels are slightly larger than the front, a local characteristic. There are some—but not many— coffee houses with terraces under gay umbrellas, a sight unusual in Eastern Europe. Children in from the countryside have starved, extraordinarily aged faces. Few, if any, Communist signs or symbols are in evidence; I never saw a hammer and sickle, or a portrait of Wladyslaw Gomulka, the party leader. True, the streets burst out with small Polish flags just before May Day, but this seemed to be a nationalist rather than a Communist demonstration, although I did see one—but just one—photograph of Lenin. Flags and portraits are also to be seen on July 22, the country's national day.

The streets are spotlessly clean, and all traces of snow, even after a heavy fall, are removed within a day, something that cannot be said for Chicago, as an example. The streetcars are colored cream-and-tomato. Pets are rare. I saw a few queues, mostly for the buses, and, on occasion, people will line up for food—in particular meat and cheese. A tree, if it becomes diseased, is at once replaced by a new adult tree, not a sapling; old, old women do this work. One curiosity is that a street called Rutkowskiego, full of shops, is closed to traffic from 1 to 7 P.M. every day, like Florida in Buenos Aires, so that people can wander down its length more easily. Another—most striking—is that all elevators in buildings throughout the city may be opened by the same key. I asked why, in that case, it was necessary to lock them at all. Reason: to keep children out.

The Vistula, which bisects the city in a straight wide stream with one loop, was first bridged in 1573. Both banks are landscaped—one bank is higher than the other—and lined with green. The river does not look particularly broad, but nobody but a madman ever tries to swim it; the current is too swift, and the banks, made of shifting sand, too treacherous.

Warsaw has, inevitably, two Chopin museums, and one devoted to Madame Curie. The Old Town Market Square, the Stare

Miasto, is well worth a visit; here the reconstruction of Warsaw attained its highest, most loving pitch. It resembles to a degree noble squares in Bruges and Siena, but is smaller. One jagged tooth of a ruin stands nearby—the remnants of the Royal Palace, blown up by the Germans. Even at this date, no final decision has been taken as to whether to reconstruct it or not; the cost will, it is thought, be prohibitive. The new opera house, to be mentioned below, is an imposing sight, and, in a different category, so is the elaborate Wilanow Palace, which was built and lived in by King John III Sobieski, who saved Vienna from the Turks in 1683. Smaller, more graceful, of the utmost charm, is the delicious little Lazienki Palace, situated on an artificial island surrounded by a pond. The legend is that it was built by an early eighteenth-century nobleman for his mistress; but he had a second mistress, a mermaid, who had to be content with the reflection of the real palace in the pool.[8]

A notable horror is the Palace of Culture and Science, which looks like a gangster's wedding cake. A gift from the Soviet Union, it represents the most old-fashioned period of Stalinist architecture. It has 30 stories, rising to 235 meters (770 feet), is built of solid marble, and contains more than 3,200 rooms; here various cultural and other associations have their headquarters, including the Polish Academy of Science. Five different theaters are housed in it as well. Poles in general detest this elaborate tricked-out structure, but it is difficult to avoid. A standard Warsaw story is that the luckiest man in the city is its watchman because, living inside, he can't see it. Another joke of former days is that the building should be called St. Joseph's Cathedral and that the spire was made high and sharp so that Stalin could the more easily be impaled on it.

The most moving sight in Warsaw is, as is natural, the ghetto, now restored and made into a housing development with rickety

---

[8] This exquisite palace was saved from destruction by the Germans just in time. The Nazis had drilled several thousand holes in its walls for dynamite charges, but they were forced out of the city before they could set them off.

tenements. It is not—and never was—the sole Warsaw area occupied by Jews, who have always been scattered throughout the
city. At the entrance is a blunt, well-executed monument to
Mordechai Anielewicz, the leader of the uprising, and his heroic
comrades. After the fighting here some 100,000 Jews were exe-
cuted by shooting, quite aside from those killed in the fighting or
murdered in the concentration camps. The ghetto area rises a
few feet higher than its surroundings, because there are an estimated 200,000 corpses underneath and it was impossible to bury
them individually. So a crust of concrete covers the whole tragic
scene.

Affairs Cultural and the Like

Warsaw has always prided itself on its intellectuality; books are
cheap and sell in large numbers. I might mention in passing that
a Polish translation of my *Inside Africa,* which was a sellout here
but in a small edition, cost only 90 zlotys ($2.25) despite its
bulk.[9] The most popular American author is, as is to be expected,
Ernest Hemingway; John Steinbeck also had a large vogue, but
this has been damaged by his hawkish stand on the Vietnam
war. Other books I noticed translated from English and French
were *Sniadanie u Tiffanyego* by Truman Capote, *Nocny Lot*
(*Night Flight*) by Antoine de Saint-Exupéry, and a collection of
Faulkner stories under the title *Roza dla Emilii.*

American publishers might learn a good deal from an ingenious Polish system which determines how big an edition of a
new book shall be; there is no problem here about returns or
remaindering. According to a recent article in the *Times Literary Supplement* of London (January 25, 1968), all publishers
formally give notice six months in advance to the central organization of publishers, the Dom Ksiazki or Book House, of what
titles they intend to put out. The Dom Ksiazki then circularizes
its members (some ninety in Warsaw), asking for an estimate of

[9] The American price was $6.50 when the book first appeared in 1955, but
is higher now.

what sales will be. These estimates are tabulated and returned to the publisher, so that he can gauge with considerable accuracy how many copies to print. There are about seventeen hundred bookshops in the whole of Poland, and Dom Ksiazki has fourteen thousand traveling salesmen to serve them, down to stalls in the most remote villages. Publishers are both state and cooperatively owned. An event of considerable importance in the book world is the annual Warsaw International Book Fair. Last year publishers from twenty-eight countries, including the United States, attended its sessions; one feature was a portable language laboratory for speeding up translations.

Contemporary Polish writing has little interest, if only because most writers are torn between socialist realism, the orthodox doctrine, and the new wave of nonhero modernism. The most distinguished living Polish writer is probably Antoni Slonimski, but he is little known in English. Two Polish winners of the Nobel Prize for literature, Henryk Sienkiewicz (*Quo Vadis*) and Wladyslaw Reymont (*The Peasants*), are still widely read. Recently a youthful Marxist philosopher, Lezsek Kolakowski, one of the best-known writers in the country, was expelled from the party and put under wraps for daring to dissent from the regime.[10] Other new talents have skipped from the country to live abroad.

Periodical literature is lively. One interesting magazine, *Stolica*, began publication in 1945, and directs itself mostly to the Polish intelligentsia; its local circulation is around sixty thousand, and, one of its editors told me, it reaches Polish readers in fifty-two countries all over the world. It is not a directly political or party organ, and concentrates on Warsaw the city and its cultural life, even printing TV programs in detail although these will be long outdated before the magazine reaches its readers abroad. There is no local variant of *Time* or *Life* (like *Ogonyok* in Moscow), but an imitation of *Reader's Digest* is successful. There are several serious literary magazines on the pattern of *Harper's* or the *Atlantic Monthly*, and a poetry magazine which

[10] "The Polish Scene," by S. L. Shneiderman, *New York Times Magazine*, April 2, 1967.

has risen from a circulation of four to ten thousand in the past few years. A woman's magazine on the style of *McCall's* or *Ladies' Home Journal*, but much less elaborately put out, has the largest circulation in the country, two million. The biggest newspaper circulation (around 400,000) belongs to a publication in Katowice, the industrial city in southwestern Poland. The biggest in Warsaw are *Life of Warsaw* and the *Evening Express*, with around 300,000 each. The official party paper, *Trybuna Ludu*, is much smaller.

There are about forty cinemas in the city, about twenty theaters. One movie showing while we were there was *Cat Ballou* with Jane Fonda. TV (one channel) runs for ten hours a day; the morning programs are mostly educational. Among plays the most successful recently was a satire called *Tango*, by Slawomir Mrozek, which skillfully laid open some of the weaknesses of contemporary Polish society; Warsaw theaters are 95 percent sold out most of the time, and we couldn't get a ticket for it when it was running. A Jewish theater, headed by Ida Kaminska, flourishes in spite of the ugly present-day wave of anti-Semitism.[11] Western visitors should remember that, as in Moscow, outer garments such as coats, hats, or galoshes are not permitted in the auditorium of any theater, and must be checked, a process which often entails a bit of queuing. Another point is that Warsaw has enough public entertainment in various fields to make possible the publication of a weekly guide devoted solely to the theater, movies, and so forth, like *Cue* or the *Semaine à Paris*.

The new opera house, called the Wielki, is the biggest in the world, and cost about $20 million. Wielki means "Great." Opened in 1965, it has three imposing colonnades, miles of shining floors alternating between parquet and marble, twenty elevators, a vast stage, a curtain which weighs twelve tons, oversize spittoons, a staff of thirteen hundred, beautiful and highly modern glass-mosaic clocks set in the walls, several three-ton chandeliers, and

[11] Madame Kaminska recently announced her intention of leaving Warsaw and emigrating to the United States and, for the time being at least, the Jewish State Theater is extinct.

enough electricity for a town of sixty thousand. It was built to replace the opera destroyed by the Nazis, and is a national monument. It reminded us to a degree of the Moscow subway, which is similarly magnificent; its purpose is not merely functional, but symbolic. Here, the Wielki seems to say, is our pledge to the future, to social enjoyment and advance, and to the concept that culture is open to everybody, without exception or favor, down to the lowest peasant. Here, too, men and women who live dull and circumscribed lives may find a touch of glamour. The repertory is a mixture of Polish works and standard classics. The opening performance was devoted to *The Haunted Manor,* a Polish work by the nineteenth-century composer Stanislaw Moniuszko, and this is often repeated. During our stay in Warsaw the repertory included two other works by Moniuszko, a Stravinsky ballet, several Polish operas, *Giselle, Faust, Carmen, Aïda,* Puccini's *Cyganeria* (*La Bohème*), and *Don Quixote,* spelled *Don Kichot.* Wagner is never played.

✿

The graphic and plastic arts flourish in Warsaw these days, and, in strict contrast to the case in Moscow, follow a highly modern trend. Most painting is altogether avant-garde. We went to a gallery called the Zacheta, among others, and saw dismembered nudes in the manner of Francis Bacon, stripes of blurred color like Rothko, drawings like those of Saul Steinberg, and such examples of pop art as a cat and a green moon under a red table, atop which is a nude with a bearded mask and dewdrops, in the form of jewels, set in her hair like nipples, next to a pink snail. Another painting showed the carcass of a butchered lamb on a trapeze with a lion and lioness holding hands, and still another portrayed a mad bull—or wolf?—raging through blue, maroon, and ocher weeds. There were several mobiles made of lath and string, and plenty of exhilarating and vivacious posters, which are a Warsaw specialty.

Politics, the Purge, and the Jews

"Poland has only two troubles, the weather and our neighbors," is a well-known Warsaw crack. I asked a dozen men and women whom the Poles hated most, the Russians or the Germans. The answer depended on whether the person questioned was a party member or not. The Germans are so universally detested that many Varsovians make little distinction between West Germany and East. *Germany* is the enemy—moreover, an enemy with a good case for vengeance, because Poland, shoved bodily to the west after World War II, possesses a large bloc of former German territory.

As to the Russians, they are regarded, in the most charitable view, as a kind of defense or guarantee against any future attack on Poland by Germany. Of course the Poles are obliged to follow Moscow—as in the brutal Czechoslovak crisis in August, 1968—if only because they are helpless before the weight of the Soviet military establishment. Poland is the only country with the Red Army on both sides—in Russia on the east, in East Germany on the west—and on its own soil as well. A small Red Army force is stationed permanently in southern Poland; its soldiers are seldom seen, however, and are kept strictly under wraps. The Czech affair may change this. To sum up, the Russians are not adored by any means, even by party members. A little joke expresses the point. Warsaw Citizen A tells Citizen B that the Russians, in their space flights, have reached the moon. "What!" exclaims Citizen B. "*All* of them?"

Wladyslaw Gomulka, the first secretary of the Politburo of the local Communist Party (called the Polish United Workers Party), is, as is known to everyone, the most powerful personage in the country—a grayish man, shy, devoid of flamboyance, a believer, even an idealist, dedicated, disciplined, and sharp of instinct. Born in 1905, he comes of peasant stock, and began adult life as a manual worker. He is a recluse, seldom seen publicly. When the Stalinists took over Poland after the war, he was accused of be-

ing a "dissident" and was imprisoned for a number of years.
Stalin proceeded to run Poland through more pliable Polish
agents. In October, 1956, this Stalinist regime reached the end
of the line. Stalin himself was dead. Gomulka, who commanded
great prestige among Poles, especially the party activists, was
released from jail and rode to power on a new wave of national
feeling and hope. He negotiated skillfully for Poland in the 1956
"insurrection" against the Russians, which led to the breakout
in Hungary. Gomulka had courage enough to defy the most
extreme pressure from Khrushchev, who surrounded Warsaw
with tanks but, at the last moment, did not let them loose. Ever
since, Gomulka has been top man in Poland. One of his first ac-
tions on assuming power was to release from jail Stefan Cardinal
Wysznyski, who had also been imprisoned by the previous
regime and who is probably today the second most important
man in Poland, and work out an agreement with him on such
touchy matters as religious instruction in the schools. But let
nobody think that Gomulka is a liberal or anything like it. He is as
hard-minded as the Kremlin, but practical, pragmatic, and, most
Warsaw citizens seem to think, a decent enough human being.
The key to his relationship with Moscow today is that, in exchange for a free hand domestically, he cooperates to the full
with the Soviet Union on foreign policy.

One of his principal predecessors in power, the pro-Stalinist
Jakub Berman, has never been molested. I saw his villa near
the American Embassy in Warsaw. The Poles do not kill, as do
the Hungarians, Bulgarians, and Yugoslavs.[12]

In the agitated spring of 1968 a difficult period began for Gomulka, after earlier rumblings. A formidable rival rose to challenge
his position—Major General Mieczyslaw Moczar, a war hero,
nonintellectual, leader of the partisans, and Minister of the Interior and chief of the secret police, who had stuck it out in
Poland as a resistance fighter during the whole course of the

[12] But 54 citizens were killed in street fighting in Poznan, in the uprising
which set off the 1956 "revolution." Flora Lewis in the *New York Times
Magazine*, June 12, 1966.

war. Gomulka, his enemies said, was becoming old, tired, and inept; the country needed more economic progress under a stronger hand. This tussle broke out into the open after serious student riots which took place in Warsaw and elsewhere in Poland in March, 1968. These in turn became involved with the oldest and most despicable of Polish bogeys, anti-Semitism. "Anti-Semitic" means more than "anti-Jewish" in Poland; it means "anti-intellectual"—the revolt of the lout. Gomulka has never been particularly anti-Semitic; his wife is Jewish. Moczar is, on the contrary, reputed to be somewhat more anti-Semitic. Then, too, another complicating factor, many members of the Polish establishment are anti-*Zionist*, even if not anti-Semitic, partly because their Russian allies and coadjutors take this position, partly because of internal Polish stresses. For a man to be a Zionist means that he is less good a Pole. Jews are not even called Jews in normal talk in Warsaw, but "Poles of Jewish descent." Much anti-Semitism also derives from postwar days when Stalin ran Poland largely through Polish Jews who had been *émigrés* in Moscow during the war. It was against this group that Gomulka rose in 1956. He himself divides the Jews today into three groups —solid Polish citizens, "whom we embrace"; those with allegiance torn between Poland and Israel; and pro-Israelis or Zionists.

Trouble began—typically enough for Warsaw—over something literary. During performances of a classic Polish play, *Dziady*, by the nineteenth-century poet Adam Mickiewicz, members of the audience laughed and cheered at anti-Russian remarks in the text. The government in consequence closed the play down. This provoked a kind of "revolt" in the powerful Writers Union and a furious students' protest at the theater early in March; about thirty boys and girls were arrested. All were released the same night, except a few so-called ringleaders, two of whom were Jewish. This action was protested by the Writers Union, other cultural organizations, and some divisions of the university. The government cracked down again (this was in Moczar's province), and several hundred young people were arrested after a demonstration. Only the names of *Jews* taken by the police were given out to the press—as if to capitalize on the riots by deliberately inciting an anti-Semitic feeling. If Gomulka made any protest, it was not heard. Then rioting spread among students, both Jews and non-Jews, throughout the country, and gave the country its biggest crisis since 1956.

The government's reply was a purge of limited scope but relentless severity. It was still going on when we visited Warsaw late in April. Four divisions of the university were shut down, and dozens of students expelled. The university was, we heard it put, "destroyed." The Jews were persecuted not as Jews, but as "Zionists," a hypocritical touch. The party boss in Silesia, an important and ambitious character named Edward Gierek, seemingly joined forces with Gomulka, but was said to be secretly on Moczar's side at the same time. Some nine thousand members of the party were expelled, a fate equivalent to political extinction, and about ninety senior officials were dismissed from their jobs, which could be the equivalent of starvation if they did not have friends or relatives to support them. Among them were the head of the Atomic Energy Commission, several high officials of the Foreign Office, three generals, the men in charge of Olympic sports, and multitudinous actors, movie directors, and show business people at large. [13]

Ninety percent of those purged were Jewish, although this was never given as the excuse; the familiar phrase was that a person had "adopted a political stand inimical to the interests of the party and the state." In a single publishing house 140 employees were dismissed. One irony is that the rank and file of ordinary *people* of Poland, Jews and non-Jews, were almost unanimously as well as passionately pro-Israel during the six-day war between Israel and the Arab states in 1967, despite the attitude of the government. Another is that only about 30,000 Jews—as against 3½ million before the war—survive in Poland today, of whom only 5,000 reside in Warsaw. Of course, anti-Semitism has been endemic in the country for a thousand years. I asked a Polish

[13] *Time*, April 12, 1968.

friend to explain why non-Jewish Poles should have such a fierce and numbing fear of a community numbering only thirty thousand among more than thirty million. "Ah," he replied, "they have all the important positions!" Such nonsensical talk was similarly heard in Berlin in the early days of Hitler.

✿

Never, to sum up, have I heard so much vicious anti-Semitic talk since Germany in the early 1930's, although this was often oblique, not personally directed against individuals. A good many purgees will, after screening, probably get their jobs back—so at least decent citizens hope. Gomulka has recently announced that the government will give exit visas to any citizens wishing to emigrate to Israel, but this pledge was largely vitiated by the complicated and onerous procedures involved. Apparently the witch-hunt phase of the terror is over. And, one should add in all fairness, nobody went to jail except students involved in actual riots, and these were soon released; nobody got killed. There is no comparison to the Nazi or Soviet terror. Many citizens of Warsaw think that the whole affair was no more than a kind of smoke screen to hide a governmental crisis. The Jews were the unfortunate scapegoats of a struggle for power between Moczar and Gomulka.

Finally, we must revert to the Soviet Union, on which the life of Poland depends. The principal thing to say is that, no matter what amenities Warsaw may have, and they are considerable, this city is the capital of a regime based in the last analysis on force —harsh, brutal force—moreover, a capital which, to make things worse, is held in thrall by another regime, that of the Soviet Union, which is even harsher, more authoritarian, and more tyrannical.

9. MOSCOW I—ITS BEAM
AND BULK

Moscow, the central bastion of the Other World, the epitome of Soviet power, has changed a good deal since my last visit twelve years ago, although much remains the same. This is a city still replete with contradictions, challenges, defiances, and loaded secrecies. It would be a great mistake to underestimate its force and weight, capacity for further development, and sheer inhuman drive—if only because if the regime it represents did not have such vitality and power, it would not be the universal worrisome preoccupation it is. One need only recall the occupation of Czechoslovakia in August, 1968.

Physically Moscow, laid out in concentric rings with arterial spokes, is dominated by a single central structure, the Kremlin, which is both the seat of government and a kind of mystical holy-of-holies. I know no other capital in the world wherein one close-knit cluster of buildings, directly in the middle of the city, so effectively symbolizes an entire regime. It looks on Red Square, another potent symbol. The rest of the city is less distinguished from a physical point of view, although many streets are unusually wide and a handful of ornate white skyscrapers soar above the ash-colored flatness of the city as a whole.

The four most clearly discernible changes in the past decade in the nonpolitical field are an improvement in the standard of living, a large increase in housing (but not large enough), the growth of automobile traffic, and new formidable complexities about money.

I am writing about Moscow the city, not the Soviet Union as a country, but in passing there should be a line about the national government, because the attitudes of this are reflected in municipal affairs. In a word, the Brezhnev-Kosygin regime is stand-pat, don't rock the boat, play-it-safe, based on the Soviet interpretation of law and order, with a large vested interest in the status quo. Khrushchev, although still alive, is as dead as Stalin. The government will strike and strike hard when it considers its vital interests to be at stake—as in the crisis over Czechoslovakia— but the period of pranks and adventurism is over.

As to improvement in the standard of living, this can most easily be seen in the way citizens are better dressed, more consumer goods are available in the shops (but not enough), and more money is being spent and saved. The standard comes nowhere near equaling ours, but the advance is striking. One evidence of this is the queues for foodstuffs and similar basic items are no longer major burdens for shoppers. People spend money on articles hitherto unknown, like sports goods, and skiing is a rage.

The advance in housing is spectacular, but a great deal still remains to be done, as will be seen in the following chapter. The drive into Moscow from the airport is startling to the visitor who, like me, is returning after a good many years away, because the boulevards are solidly lined today with large new housing developments in what was virtually open country in the 1950's.

As to traffic, Moscow has, believe it or not, entered the automobile age. It doesn't have traffic jams or serious parking problems yet, but there are plenty of cars in the city (about 150,000 at the moment), and the hiss and gurgle of traffic outside the Hotel National, in the center of town, can keep a person awake at night. We drove out to Zagorsk, thirty-three miles away, to see its celebrated monastery, and our small Russian-made car—a Volga, equivalent to a Citroën—had to thread its way in and out of solid lines of trucks for the whole route. Moreover, passenger traffic will soon increase substantially following an agreement with Fiat for the production of 660,000 cars a year beginning in 1969. About half of these will be assigned to Moscow.

The growth of automobile traffic has seemingly caught the planning authorities by surprise; few new housing developments have garages. Traffic is strictly controlled; twice we committed minor derelictions (once with an American friend, once with our Intourist guide), but a cop let us off each time, after argument. Actual accidents are a serious matter; anybody involved in more than three automatically loses his driving license for life. As a safety precaution only two persons are allowed in the front seat of a car. The increase of traffic has produced baffling minor problems, some in the realm of comedy. An American official's car broke down recently on a major boulevard; it couldn't be moved, but parking was forbidden here. What to do? The resulting argument with the literal-minded policeman took an hour. Driving licenses are issued to Muscovites only after prolonged and searching examination, and to be a professional chauffeur is a big thing. A driver must know how to take a whole car apart and put it together again. A chauffeur's basic wage is 165 rubles a month (about $175 at the official rate of exchange), a substantial sum for Russia.[1]

There has been a certain amount of loosening up—not much, but some—in such fields as censorship, surveillance, and the like. Consider such a factor as that no fewer than 1,800,000 tourists, including about 25,000 Americans, visited the Soviet Union last year, most of whom spent at least part of their time in Moscow. Of course many of these visitors came from Eastern Europe or were members of Communist delegations from other countries abroad. Tourists from the bourgeois West are particularly welcomed, if only because they bring in foreign currency, for which the Soviet need is large.

But the point to make is that no secret police, no bureaucracy, is capable of carrying out close surveillance of such an immense

[1] Officially one ruble equals $1.11, but the actual value of the ruble is much less. One dollar = 90 kopecks (official rate).

number of people on top of its other duties. Hence, surveillance of the casual visitor is at the minimum—or so it seems at least. Bags are practically never opened at the frontiers or airports. The customs examination is much less exacting than in the United States. It is, however, difficult to be categorical on this general theme. Russia is a notoriously complicated country, and there are Noes to every Yes, Yeses to every No.

My mail was certainly opened (very clumsily, too), but I do not think that our rooms were ever searched or that we were ever followed on the streets. As a matter of fact, there was no need to follow us, because the Intourist apparatus generally (but not always) knew exactly where we were going. Perhaps our room was bugged, perhaps not. In any case we had nothing to hide, and said anything we felt like saying. But *if* a person is somebody whom the authorities do seriously want to watch, there is no limit to the amount of endeavor that may be applied.

It is still undeniably difficult to make contact with Russian citizens (not because they are unfriendly but because the government does its best to keep contact with foreigners at the minimum), but I met more and had easier communication with them than on any of my four previous trips to the Soviet Union.

But Moscow is still, it should always be remembered, the capital of a *closed* society, even though a few tiny apertures have been opened in the curtain. For instance the London *Times, Le Monde* of Paris, a Swiss newspaper, and the International edition of the New York *Herald Tribune* are now on sale—but in restricted numbers under the counter and only to foreigners in the hotels. This is an advance, but it doesn't mean much so far as the rank and file of the population is concerned. The BBC and Voice of America are no longer jammed in the ordinary course of events, but jamming was resumed during the Czechoslovak crisis. One reason why so many Russian children study English is to be able to listen to American and British broadcasts. A remarkable phenomenon is that, although most Moscow citizens are as isolated from contact with the external world as if they lived on Mars, they have sharp knowledge of what is going on. Of course, there are gaps. I met one young man who, in a discussion on American politics, said that Nelson Rockefeller, if elected President, would be the first Jew ever to hold that office; he had assumed that Rockefeller must be Jewish because he was so rich.

As to external political affairs in general, the people at large are becoming increasingly wary of propaganda, their own and others'. The local press, radio, and TV are equally slanted, and news is not merely distorted, as everybody knows, but deliberately suppressed. The Soviet rulers simply will not risk letting their people know the other side on any political issue that is even remotely sensitive, because of what has been called "fear of contagion."

Nevertheless there have been substantial changes, most observers would agree, in this general area within the framework of Soviet society. The regime is oppressive, yes, but people talk more freely. The scientists, engineers, technicians, have more contact with the West than the rest of society because their work makes this necessary; they can read what is forbidden to others, and, in a vast invisible network, they pass the word along. Abstract art, psychoanalysis—taboo subjects for years—are discussed. The painter Marc Chagall, who fled Russia many years ago, would be welcomed back today, if only because of his prestige. Science fiction is a craze because readers interpret the description of such things as a new social order on distant imaginary planets in terms of what might be going on in Moscow. The government has a monopoly on publishing, but a widespread illicit press exists called Samizdat, printed in mimeograph form and distributed from hand to hand. Most citizens are, it would appear, not against the regime *as such,* but, like people in the satellite states, they want to reform it; they say that they are for "good" Communism, and want the chance to vote for a choice of candidates even if all stand within the party. There is more mobility within the country than formerly, and thus more opportunity for exchanges of view. In a word, a good deal of intellectual fermentation is going on under the surface, centering in Moscow.

Power, confidence, vitality, durability—all these characteristics may be found in Moscow, and in some ways it is probably the most exciting city in the world even though much of its atmosphere remains dreary, joyless, dour. One reason for joylessness is that this is a regime built upon sacrifice. The present generation is made to suffer for the hope of the future. Also, it must be kept in mind, as far as the capital is concerned, that *Russians* do not find Moscow dull or dreary. The great capital is, indeed, Mecca to citizens by the millions in the outer spaces of the country—a brilliant shrine, lodestone, magnet. They feel, moreover, that it *belongs* to them; that it is the supreme glittering symbol not merely of power but of opportunity. And this is a city dearly loved. The poet Mayakovsky said once, "I would like to have lived and died in Paris, if there were not a universe called Moscow." Napoleon wrote before his invasion: "If I occupy Kiev I shall have taken Russia by her legs; if I occupy St. Petersburg I shall have taken her by the head; but by occupying Moscow I shall have stricken her in the heart."

Some Mechanics of Arrival

One of the basic laws of travel these days is that almost all metropolitan airports look alike, and Sheremetyevo, where my wife and I arrived from Warsaw, is no exception. It is new, smart, well kept up, and handles flights to and from the West. Some signs in English are not quite idiomatic, for instance: "LEFT LUGGAGE ROOM." Outside, in the familiar manner of Communist countries, are placards of modest size in English: "MOSCOW—A HERO CITY," "HAPPY JOURNEY THROUGH THE URSS" [*sic*], and "BY AEROFLOT TO ALL COUNTRIES." Aeroflot, the Soviet airline, is commonly described in Moscow as the biggest in the world, and, by a count of passengers carried, it probably is.[2] Sheremetyevo was named for a wealthy aristocratic family (the richest in Russia next to the royal family before the Revolution—one of its mem-

[2] An estimated 48 million people flew on Aeroflot in 1966. *Business Week*, April 2, 1967.

bers owned 825,000 hectares and 210,000 serfs) which had a large property here. A single sight in Moscow today is the principal town palace of the Sheremetyevs, now a museum.[3] The airport is one of four large airports on the perimeter of the capital. Domodedovo (Grandfather's House), twenty-five miles southeast of the center of the city, handles traffic to Central Asia and Siberia. Vnukovo deals with flights to other parts of the country, as does Bykovo. There are also several military fields. In an effort to eliminate aircraft noise, all are situated in such a way and far enough out of town so that very few planes ever pass over Moscow itself. Yet all are within a fifty-minute drive to the center of the city. Stacking is unknown.

Procedures for the passenger on arrival at Sheremetyevo are conventional enough. A single inspector deals with passports, health certificates, and customs.[4] Ours looked aghast, but in a friendly way, when he saw that my wife and I had nine bags. "All for *you*?" he asked in English. I said that we were carrying a lot of books. "Books in *Russian*?" I said no, and he passed us without opening anything after a sharp blue-eyed look which said, as clearly as words, "Can I trust you?" All this marks a considerable change from the time, some twenty years ago, as I remember well, when I saw a customs inspector slit open a tube of toothpaste a traveler was carrying, to see if it contained secret papers or contraband. Nowadays one minor formality does remain—you have to declare what money you have with you, also jewelry. But this is to ensure being able to take your valuables out freely when leaving the country. Every traveler is given a slip on which, when you cash money or traveler's checks, the bank or shop is supposed to make note of the transaction. However, nobody bothered to pay much attention to our slip when we left the country.

The Intourist apparatus took us over after we passed through

[3] It has a private theater equipped with a machine for making thunder, the only one I have ever seen.
[4] This convenient innovation was adopted lately by the John F. Kennedy Airport in New York.

customs; there was the usual snafu checking our names (in English) with their Russian transliterations on the reservation list. The Intourist man insisted that he had ordered a car too small to hold us and all our baggage, and that we must wait for another to be sent out from town; we said that we would be glad to take the small car and hold our suitcases on our laps, but this, we were instructed, would be "uncultured" and would not do. The ensuing delay cost us half an hour. Driving into town—it was about 10 P.M.—I began to feel some appreciation of how Moscow had changed. A broad boulevard, which did not exist ten years ago, leads straight into the city; there was only one red light in twenty miles. I have already mentioned the new buildings, housing developments, lining most of this route. I noticed shiny new white booths at the bus stops, mostly empty (ten o'clock is quite late for Moscow); cranes everywhere, patches of rough dirt in the grass, like bunkers on a golf course, to mark excavation sites; window displays in shops on the edge of town, which would have been unheard of a decade ago; decorations for the forthcoming May Day made of lines of single bulbs, some in the shape of a spray of flowers, not tubes of neon, which made them more effective; above all a sense of drive, push, sweep, pride, magnetism, power. My wife and I had the same thought at the same instant: "Poor Poland!"

We put up at the National. This is one of the most remarkable hotels on earth. It has housed an astonishing number of VIP's from a variety of countries for many years. Lenin lived here for some months (Suite 107) a half century ago and harangued the crowds from his balcony; a plaque downstairs commemorates his stay. My wife and I had No. 115 this time, and we were told that recent occupants of this comfortable corner room with lavender walls had been Sol Hurok and Harry Belafonte. The National rises in the heart of the area destroyed by fire in the time of Napoleon, and faces the chrome-yellow walls of several buildings within the Kremlin; from some windows you can see Red Square and catch obliquely a glimpse of St. Basil's, the cathedral with its celebrated ornate tower and turnip domes. Diagonally opposite the hotel, on Revolution Square, once stood the Temple of the Iberian Virgin, on which the early Bolsheviks hung their defiant poster, "RELIGION IS THE OPIUM OF THE PEOPLE," but this has been torn down. Next door is the liberal arts wing of the university, painted a pale yellow, and nearby are the Czar's former stables, now an exhibition hall, in a yellow paler still. Yellow and red are the predominant colors all around.

The National looks shabby and, as a matter of fact, is—although it is the Moscow equivalent of Claridge's in London or the St. Regis in New York. Two pots of frayed, half-dead pansies surround a narrowish door under a portico shaped like a piano lid. The lobby, with a newsstand at one end, has a high ceiling, but is no bigger than a smallish shop; it contains three hard straight chairs, no more. There are two elevators, behind immensely heavy brassbound scrolled doors, facing one another under arches of white plaster, and held up by heroic nude statues (but with fig leaves) of male athletes. The elevator operators are oldish men in well-worn brown uniforms; cigarettes dangle from their lips even when they are on duty. Sometimes the elevators work.

Even though it is a bit run-down, the National has a wonderfully romantic old-style atmosphere. Here—clearly—life has been lived. I did not even mind the stout women who, sitting squatly between the elevator doors on each floor, hand you messages and dole out your keys; they are the equivalent of a concierge, and of course keep tab on all your doings. Most have stainless-steel or solid-silver rows of teeth.

Rooms are identified by a number painted in large silver digits on a door made partly of frosted glass. There is only one push button for service, and ours didn't work; to order breakfast I had to go out into the hall in a bathrobe every morning and search for a servant. When I asked if the push button would ever be repaired, the answer was a placid "Perhaps no." We did not have a grand piano (most luxury suites do), but we were the possessors of an immense refrigerator. Next to it, as is universal in Russia, a stout clothes brush hangs on the wall; why these are not stolen I do not know, but I never saw a room in a Russian hotel without one. This is a country much beset with the desire for neatness. (You can be arrested and fined for dropping a cigarette butt on a street.) Dominating our living room were, first, a carved dark wood desk at least nine feet long, with a shiny polish, that must have weighed a ton (the drawers wouldn't open) and a circular table with a heavy damask *fringed* cover. The fringe is never missing. Serving a meal, the waiter invariably flips up exactly half the fringed cover; on the other half your reading matter, things purchased in shops, et cetera, remain undisturbed. We had a very large radio, which I was never able to work, but no TV. Out in the hall, however, there are numerous alcoves with TV sets, where residents of the hotel may sit and watch. Our bathtub had a plug. A friend accompanying us in a nearby room did not, however, have a plug in her tub, and could bathe only by slopping up water from the outlet with her hands.

Narzan, the admirable local mineral water, was sometimes served free, sometimes charged for (about 20 cents a bottle); there didn't seem to be any rule. The bottles are capped now like soft-drink bottles in the United States, but are apt to be unevenly filled; in the old days the cap was a cork bound with rusty wire. If you want service, there will always be an English-speaking operator available at the Intourist Service Bureau downstairs. (How many Russian-speaking hotel attendants are there in the United States?) The toilet paper is corrosive, but laundry comes back in half a day, still warm like French bread out of an oven.

A word about other hotels. New ones are going up all the time, but hotel space is still inadequate; long queues of travelers wait in the lobbies, and citizens sleep on sofas. One hotel just completed, but not quite operating fully as of the time of our visit, is the Rossiya—an immense white structure facing Red Square, with six thousand rooms, which make it the largest hotel in Europe. It seems at first glance to be as smart, new, and polished as the National is aged, quirky, and full of charm, but in actuality, although well designed for functional use, its construction is crude and the rooms are small. Intourist hotels, reserved exclusively for foreigners, are the National, Metropol, and Berlin, which was formerly called the Savoy; all are within a few minutes of Red Square.

Hotels predominantly for Russians are, among others, the Leningradskaya (a gingerbread skyscraper in the "best" Stalin-Gothic manner, with miles of slippery marble corridors), the Varshava, the Ukraina, the Sovietskaya, the Moskva, and the Minsk. These are scattered throughout the city. Several were formerly used to house diplomats waiting for permanent quarters; several take in foreigners as well as Russians in the summer season, when the Intourist hotels are full to bursting. An equivalent of Blair House exists for official guests of the highest category, and dachas (villas) in the nearby country are also made available to foreign bigwigs if they are big enough. The Minsk, on Gorkova Street, was built on the site of the old Lux, a hotel guarded like Fort Knox in former days because it was here that officials of the Third International were put up as well as other distinguished foreign Communists, trailing their veils of secrecy. Now the Minsk is as promiscuously commercial as the best Hilton, and is the only hotel I saw in Russia with neon lights giving its name in English.

Matters of Food and Drink

The National has several restaurants, of which the principal is a moderate-sized room on the first floor (our second), with tables seating four to six people; part of the bizarre unexpectedness of Moscow derives from the fact that, if a restaurant is crowded, as most of them always are, other diners may be seated at your table willy-nilly. The system makes for communication, or a try at it, despite the obvious language difficulties. We had dinners with Finns, a tantalizingly enigmatic Polish couple, and even some Russians from Soviet Asia wearing Uzbek embroidered caps. At night the equivalent of a torch singer, with a three-piece orchestra, drowns out most conversation. This restaurant serves guests who have either rubles or Intourist vouchers purchased abroad, i.e., almost anybody. The best things to eat are the caviar (although we thought that this was of less good quality than during our last visit)[5] and the marvelous Russian soups, like *solyanka,* meals in themselves. The menus are somewhat monotonous, and the service mortally slow. It is impossible to hurry the stolid, stalwart waitresses.

Russian dark bread, a kind of soft pumpernickel, is delicious, and the ice cream, which comes in any number of varieties, is usually good. The situation in regard to vegetables reminded me of the days when Henry Ford said that a customer could buy a car of any color provided it was black; similarly you can have any vegetable in Moscow provided it is peas (or fresh cucumbers in season). Nor did I ever encounter any kind of jam except strawberry. The tea is excellent, and the coffee execrable. Chocolate is rich. A kind of grape juice replaces orange or grapefruit juice at breakfast, but canned orange juice (imported from Cyprus) is available in the shops, as is mango juice from India. Patterns in food are changing, Russian friends told us. It is difficult for a housewife to find really good black bread nowadays, because white bread, considered to be more fashionable, gets priority at the bakeries. Visitors who like good old-fashioned Russian dishes like *kasha* will find these difficult to obtain, because the authorities think they are "uncultured." The government tries to push artichokes, asparagus, and sea food (in particular octopus and squid from the Pacific), but with little success so far.

The most fashionable cigarette is named Stewardess. Novotnis are an imitation of the French Gauloises, and Youths resemble standard American filter brands. The traditional Russian cigarette with its long cardboard tube has lost much of its vogue. To our ear a good many Russian names for articles seem odd. A kind of petit beurre is called "Saxophone." And, as always, I was fascinated by the free use of English. The basic Intourist menu carries a heading, "To order tasting please apply to hotel service bureau," and the sign to indicate that a table is reserved is a glass tablet decisively marked "Stop!"—a word which has been in-

---

[5] Nowadays caviar is apt to be pasteurized so that it will keep, and this diminishes the flavor. A generous portion costs about $1.60.

corporated into the Russian language and expanded in meaning.

These are some items from the English menu at the National, with prices in rubles. Add 10 percent for dollar equivalents at par:

Ham with grained cheese	1.10
Moscow borscht with a curds tart	.97
Crabs braised in a pot	1.19
Beef stewed with culinary herbs	.80
Fried pork cuts	.45
Beef stroganoff	1.44
Roast hazel grouse	2.12
Omelette surprise	1.04

Vodka is charged for by weight, with a 100-gram vial (about three normal drinks) costing between 90 kopecks and 1.04 rubles, depending on the quality. A variety known as Starka, brownish in color, is invigorating. Beer (quite good) is 30-35 kopecks a bottle, Soviet champagne 3.38-3.90 rubles a bottle, and imported cognac 5.20 to 12.75 rubles per drink. Thus a single drink of French brandy can cost the equivalent of $14.02 at par.

Next door to this main restaurant in the National is another, smaller (only eight tables), fancier, and with faster service, where the food is identical or nearly so, but which does business only in dollars or other *valuta*, foreign exchange. Rubles are not accepted. A sign warns you, somewhat unidiomatically, "GUESTS ARE SERVED FOR FOREIGN CURRENCY." Here stands a tall icebox, holding perishable goods, and at a curved table is an array of delicacies and souvenirs—American cigarettes, Soviet dolls, British gin, chocolate bars, cans of tuna fish. A chocolate bar costs the same as a Scotch—$1. A meal here will come to about $8 per person without drinks, payable in dollars or other foreign money. One legend is that this restaurant differs from its Intourist twin only in that the sugar here is superior, because it actually dissolves. "Nonsoluble sugar," which really means sugar slow to dissolve (imported from Cuba), is to be found occasionally in the Intourist restaurants, but I never encountered it.

Across the corridor, still on the first floor of the National, is something that came near to lifting the eyebrows off my head—a room with stools and a stand-up bar. Such an innovation would have been inconceivable in the Soviet Union ten or even five years ago. A stout maiden deals out Scotch (real Scotch) and a primitive variety of other imported drinks. You pay in dollars or other *valuta,* and get change in a bewildering variety of foreign moneys—sterling, Dutch gulden, French francs, Finnish marks. Downstairs there is a large *valuta* café as well. Like the bar, it is almost always packed. We even saw a few hippies—boys with beards and long hair. The bar is virtually the only place in Moscow (except a kind of night club in the Metropol) which stays open till 2 A.M. The motive is, of course, to please foreign tourists and suck dollars in.

☼

There are any number of cafés, cafeterias, canteens, in Moscow, which, according to official statistics, can "simultaneously accept" three million visitors, but comparatively few actual restaurants. It is indeed extraordinary that in this city of more than six million people there are only about twenty real restaurants outside the hotels; most are named for cities in the constituent republics of the Soviet Union or its satellites—Baku, Praga, and so on—and specialize in their native cuisines. The best is probably the Aragvi, which is Georgian. In the Baku we saw guests dance on the carpet. These are all ruble restaurants, patronized in the main by Russians who can afford them—Intourist vouchers and *valuta* are unknown. But they are extremely expensive, which is one reason why they are so few in number; the average citizen simply cannot afford to dine out. One night we drove with an American friend to a new, dashing, and expensive establishment near the Arkhangelskoye estate about fifteen miles from Moscow. Scooting ahead of us we saw a long, sleek, black limousine, with curtains drawn over the rear window, carrying license plate 1111—this could have been the car of the Premier. I reflected on what a contrast this made to circumstances of Stalin's day, when every inch of such a road would have been heavily patrolled. Several members of the Politburo have dachas in the Arkhangelskoye district, which also houses a workers' sanatorium. The area is supposed to be out of bounds to foreigners, and we had no business being there; but nobody made trouble or stopped us, although a cop made us park far down the road. We were the only non-Russians in the place. I had elk as my main course at dinner, which I had never had before; it tastes like tongue. A jukebox ("Made in Poland") played at one end of the large room. I will not forget two things: (a) dinner for four cost about $80 in rubles, and (b) my wife slipped in the total darkness surrounding this festive establishment, as we clambered through the woods toward the road, and broke a rib.

Arkhangelskoye once belonged to the Yusupov family. Its first orchards were laid out three hundred years ago, and one Prince Yusupov owned 350,000 serfs.

Some Madnesses About Money

Most of the hotels, I soon found out, maintain what are called *valuta* shops. The sign in our lobby read: "SOUVENIR SHOP IS IN THE SERVICE BUREAU SALE [*sic*] IN FOREIGN CURRENCY." There are similar establishments elsewhere in the city—which sell food, watches, cameras, furs, and so on—only for dollars or other foreign currency. This seemed to me at first as strange as if Saks Fifth Avenue operated only in rubles. The country appears to be dollar-mad. Recently a foreign ambassador received a bill for services in connection with his official airplane—$400, to be paid in dollars. Another had several rooms in his embassy painted—the bill was $1,100, due in dollars. Rubles were not even mentioned.

We happened once to share a taxi with a Russian. He did not speak much English, but as far as I could understand him he was offering to sell us a cameo ring he wore for $15. On the night of May Day we ventured out into Red Square, where large, solid throngs had congregated and were milling about. Red Square has an intimate call for nearly every Moscow heart. I saw scenes that would have been difficult to imagine ten years ago—shifting circles of lively youths surrounding a guitar player, girls and boys dancing the twist. Not a uniformed cop in sight—with a mob of at least ten thousand people churning up and down the square.

A young Muscovite approached us.

"You are American?" he asked in English.

I said that we were.

"You come from New York, Chicago, Minneapolis?"

"New York."

The young man paused. "My name is Fred," he said cautiously.

"Well, hello, Fred. Are you a student?"

"Not exactly. I am working with Tom. Please, will you meet my friend Tom?"

We were led to Tom in a dark spot on the edge of the vast, swarming crowd. Tom looked us over, and said, "I offer you 4.15."

This meant that he would give us four rubles fifteen kopecks for a dollar instead of the official rate. We would have gained approximately fourfold, but we refused because we didn't want to do anything illegal.

Similar episodes occurred several times while we were in Moscow. The phenomenon has explanations on two levels. The *government* wants dollars, which accounts for the existence of the *valuta* restaurants and shops, operating on a legal basis, in order to amass foreign exchange with which to pay for such imports as elaborate electronic instruments and the like. The Russians could manufacture these themselves, but it is cheaper and more convenient to purchase them abroad. *Private citizens* want dollars because they know that they can buy better things cheaper in the *valuta* shops than in the ruble stores, as well as merchandise otherwise unobtainable. Plenty of Moscow citizens have learned to like Scotch whiskey and American cigarettes, which can only be procured for foreign currency. So they are prepared to sell rubles for dollars at an illegal rate.

Shops are of two types in Moscow. The *valuta* shops, called "Beriozka (Birch Tree) Stores," are substantially cheaper than the ruble shops, better run and, naturally, less crowded. But I must not fall into the temptation of giving unwarranted importance to these shops, since they are comparatively few in number. In one, a food shop crudely decked out by Western standards, but smart for the Soviet Union, a variety of signs gives education both to foreign and domestic customers, like: "STURGEONS MAKE EXCELLENT COLD APPETIZERS AND SUPERB SECOND COURSES" and "THE SOFT RED CAVIAR MADE IN THE USSR IS CONSIDERED THE BEST IN THE WORLD." Over a tempting array of chocolates ("Fine and Original Taste—Plenty of Calories") was the alluring notice: "TRY OUR CHOCOLATE AND YOU WILL SEE THAT SOVIET CONFECTIONERS ARE DOING A GREAT JOB!" Although some prices are marked in rubles, nothing but foreign currency is accepted. A bottle of "milk, half fatty" costs 10 kopecks, a dozen eggs 52; tinned cream was 31, and "melted butter" 68; cold boneless smoked pork was 2.14 rubles a kilo. Among articles priced only in dollars were Armenian wine ($3.89 to $4.05 a bottle) and Soviet champagne (at $5.15). But a bottle of the best Scotch Scotch costs only $2.70, Beefeater Gin $2.20, and American cigarettes $3.10 a carton, much less than in the United States. In blunt fact, the low prices for certain goods purchased in Moscow for foreign currency constitute a kind of bribe to get dollars. For instance, one friend of mine bought a small Soviet car without delay for $1,400 in American cash; paying in rubles he would have had to put up the equivalent of $4,900 and wait two years. A bicycle can be had for as little as $15.

The Beriozkas in the hotels also feature Soviet-produced merchandise as much as possible. Here are to be found Soviet watches, cameras, confectionery, and perfumes—well-known local brands are Goal, Dajna, Dsintars, and Jolanta. The salespeople, working with small computers and adding machines instead of the old abacus, are generally agreeable, but sometimes a good deal of red tape is attached to a sale. Girls will say apologetically, "Little speak English." Lacquered Russian boxes, a well-known specialty, cost from $8 to $155.67, and a set of painted wooden dolls as much as $48.34. A guitar costs $3.75, a Shostakovich record $3.33, a fur hat around $20, and a silver fox $47.

As to the ruble shops, they are dense with customers. Here are some prices I noted at GUM, the large department store, which is really a series of bazaars under a glass roof, facing the Kremlin. Umbrellas cost 4-8 rubles; a string of coral beads, 3.80; a sheet, 4; a cake of soap, 30 kopecks; a toy train, 4.90 rubles; an "Oriental" rug, 84.20; a hunting rifle, 43.50;[6] a TV set, 100 to 1,000. Most of the textile goods seemed shoddy. Cheap shopping bags made of net are provided, in contrast to the colorful paper bags at the *valuta* shops, and sales on credit are permitted subject to authorization from the customer's trade union or employer, a new development.

Then I went to Gastronome No. 1, the principal food shop in the center of downtown Moscow. Bologna sausage cost 2 rubles (say, $2.20) a kilo; a can of sweetened condensed milk, 77 kopecks; a loaf of bread, anywhere from 13 to 28 kopecks, depending on the variety; apples, 1.50 rubles per kilo; fresh lemons, 35 kopecks each (frost in the Caucasus damaged the citrus crop last winter); cheeses, 2.20 to 3.90 rubles per kilo; turkey, 3.20 a kilo. Food is, in a word, very expensive. One peculiarly expensive item is cut flowers; roses flown from London are cheaper than the local product. A single carnation costs 1.30 rubles, almost a dollar and a half. On the streets flowers are sold singly wrapped in small cones of cellophane.

In theory, a person is supposed to buy only for his own or his family's use; otherwise the process is "speculation," a serious crime. However, the authorities encourage the sale of luxuries on the ground that, as the phrase is, "nothing is too good for the proletariat." The government actually spent about £1,150,000 in England two years ago to import men's clothing, so that citi-

✿

---

[6] Guns—even for sport—may not be sold except under license, a procedure that the United States might well copy.

zens would look smarter during the celebrations attending the fiftieth anniversary of Soviet power.

Where does all the money come from? Rubles seem to be plentiful even though wages and salaries are, by our standards, low. The average worker has a short working week, just over forty hours, but his average wage is only about 100 rubles a month. The legal minimum wage is, however, only 60 rubles. A schoolteacher's average monthly wage is 140-150. The salary of such a specialist as a movie cameraman is 200-250 rubles a month, but he will receive a bonus up to 5,000 if the picture he is working on is successful. The average engineer gets 200-250 rubles a month, a foreman 400, and a superintendent 600. The salaries of cabinet ministers and high military officers are modest in theory, but they have almost anything they want in perquisites. Some aged women work as attendants in the museums in order to supplement their meager pensions.

How—considering the comparatively low wage scale as a whole —are citizens able to spend as much money (in rubles) as they do? The queues for taxis are longer than those for buses. A customer fairly has to fight his way into the great department stores and Gastronomes. There are several reasons:

1. Consumer goods are still scant, although, as noted at the beginning of this chapter, more merchandise is available than, say, five years ago. But shortages are still severe in a spasmodic, capricious sort of way, largely as the result of bad distribution. One present acute shortage is in flatirons. So people cannot buy, and rubles pile up. The items people save for most are dachas in the country, dishwashers, TV sets, and cars. Savings accounts do a lively business, and so—a comparatively recent development—do insurance companies. Savings accounts pay a maximum of 3 percent interest.

2. Almost all families are two-earner families.

3. Medical care is free, and old age is provided for. Unemployment is unknown, according to the government.

4. Taxes are low—the maximum income tax is 13 percent. But

a "bachelor tax" annoys many youthful citizens, both men and women.

5. Rent is very cheap—never more than 6 percent of the income of the head of the household.

6. Travel and entertainment cost comparatively little.

May Day

We arrived in Moscow just in time for the May Day celebrations. It was a race that morning to penetrate the throngs and get to Red Square on time. We used pedestrian underpassages and channels. How my wife was able to find our way I do not know. In a quarter of a mile we had to go through seven different police inspections to reach our places in the grandstand.[7] Guards with rifles stood on the roofs of nearby buildings. Such precautions are, of course, justified, because the entire ruling body of the Soviet Union stands exposed for hour after hour—easy targets —and the crowd is as big and heterogeneous as at a European football game.

This celebration is, as everybody knows, one of the formidable sights of the world. The day was wonderfully brisk and sunny, with a clean wind making the multitudinous banners flap. All night before in the National we had heard the crunch of tanks and armored cars in the streets, the trial roars of loudspeakers, and the heavy flap-flap of the vertical scarlet banners, made of some stiff material, hung row by row on the façade of the hotel. The ceremony began at 10 A.M. on the split second. Two sleek gray open cars appeared, Chaikas (Seagulls), with large white tires; these never go on public sale, but are manufactured exclu-

---

[7] Our own situation was somewhat complicated because we had two pairs of tickets, one from Intourist, the other from a friendly embassy. The latter were better and so we sought to make use of them. But guards, after looking them over carefully, would not let us into the journalists' section because the tickets said diplomats; then other guards would not let us into the diplomats' section because, after scrutiny of our passports, they saw that we were not diplomats. This was bureaucracy with a vengeance, we thought. Then at a third try, another guard threw up his hands, laughed, and waved us in with the journalists. I cannot emphasize strongly enough how friendly everybody was—even though we were Americans, enemies. One man risked his own seat to escort us all the way across the square.

sively for high personnel in the military and government. In one stood Marshal of the Soviet Union Andrei A. Grechko, Minister of Defense; in the other General Eugene Ivanovsky, the Deputy Commander of the Moscow Garrison, both in blue-green uniforms. The cars separated delicately, came together, circled deftly, almost pirouetted—it could have been a ballet. Then Grechko spoke. By tradition this is *his* day, the day of the chief of the Soviet armed forces. Then a twenty-one-gun salute sounded with a reverberating bark and rumble, and the parade began.

First came the highest officers of the army, with gold belts around their jackets and broad scarlet stripes down the trousers. Most were as fat as pork, so much so that it seemed impossible that they should be able to march so briskly. But I was told that they were even fatter until the days of Marshal Zhukov after World War II; Zhukov, no sylph himself, insisted on making his senior officers slim down. (Incidentally, the two leading participants rode horseback instead of using Chaikas until ten years or so ago, when horses had to be given up because Marshal Bulganin, the Minister of Defense at the time, did not ride well enough.)

So many people all over the world have seen this parade on TV that there is no point in describing it. A novelty this year consisted of parachutists in mauve berets—also submarine personnel who marched with what was almost a goosestep. About thirty different elements of the Soviet military establishment were to be seen. At the end came the weapon-carriers with their giant silver rockets, like fat pencils. There is no fly-over or air show nowadays because the weather is apt to be treacherous and nobody wants an accident so close to Lenin's tomb and the Kremlin. When the military finished, the turn came of unending civilian delegations. Some carried banners and floats protesting against the war in Vietnam; several were fiercely anti-American, including one which said "USA" in huge letters, with the "U" a bag of chains, the "S" a dollar sign, and the "A" an atomic bomb. Another portrayed the United States as a scorpion. At the end we sought to return to the National, but it was impossible to cross the solid, unending lines of marchers; we only succeeded finally in breaking through by marching briefly with a delegation!

The parade was still going full blast at 4 P.M., although nobody participated except residents of Moscow; even so, the crowds were enormous. What the show was intended to demonstrate was, as always, Soviet power—to give citizens psychological confidence in the armed might of their officers, troops, and terrifying tanks and other equipment.

It was interesting to reflect that the institution of May Day, celebrated with such avidity in Communist countries, originated in the United States as a memorial to the Haymarket Massacre in Chicago in 1886. The Second International in Copenhagen subsequently chose and named it as an international labor holiday.

It was a surprise to discover that in hard-working Moscow the May Day festivities last four whole days, during which all normal pursuits stop. However, the Sunday of that week is considered officially to be the previous Friday (a nice little Soviet touch) so that workers do not have their holiday (Wednesday –Thursday–Friday–Saturday) interrupted; work begins again on Sunday. Then on May 9 comes another three-day holiday (non-interrupted), the equivalent of our V-E Day, to celebrate the defeat of Germany in World War II. There was little chance to do much work in Moscow that festive week.

Facts, Figures, Fundamentals

Moscow is a double capital, both of the Soviet Union as a whole and of its principal constituent republic, the Russian Soviet Federated Socialist Republic (R.S.F.S.R.). With about 6,500,000 people, it is the sixth largest city in the world from the point of view of population. The figures have risen steeply since the Bolshevik Revolution. The metropolis had about 200,000 people in the time of Napoleon, an even million in 1900, and roughly 1,850,000 when World War I broke out in 1914. A very large industrial city, it contains about a million workers; one plant alone (automobiles) employs more than 70,000. As the principal educational center in the Soviet Union, the city holds some 600,000 students. Resolute and systematic efforts are being made to check its voracious growth. The authorities do not want it to become bigger, if only because it will become too unwieldy to handle. Housing is enough of a headache already. Population is allowed to increase only by natural growth, and, if the present ratio of births to deaths carries on, the number of people will rise by only 300,000 by 1980. Measures severe in the extreme are being taken to check any increase of population by other causes; permits to move into the city are difficult to get, even if this means that a family may not join a worker there, and is split in consequence. It is even more difficult to procure authority for a new industrial enterprise or institution. As a palliative measure satellite cities, like those in Paris and London, are being planned for the outskirts. For a person to enter Moscow and live here for any time illegally is virtually impossible, because of surveillance and the constant need to show an identity card, even for such a trivial act as picking up a package at the post office. One form of control over old people is the address to which their pension checks are sent. There's no room for grandma any more.

Moscow, as I have already mentioned in passing, is built in concentric rings around the core of the Kremlin; it covers 347 square miles, which makes it the fourth biggest city in the world in area. A circular highway some sixty miles long defines the city limits about ten miles from the center of town. Radial "spokes" intersect the several rings. Some important Moscow streets, like Tchaikovsky Street on which the American Embassy stands, are immensely broad by Western standards, and can be three times wider than Park Avenue. Beyond the exterior boulevards is a "forest belt" (covering some 445,000 acres) much like the Green Belt in London, except that the trees here are mostly pines and birches. The city itself plants 400,000 trees a year, and, Soviet statistics tell us soberly, has 2,500,000 bushes. There are 400 parks (as against 34 in 1917); one of these covers 2,832 acres, and is by far the largest in Europe. There are about 2,000 streets (some authorities say more); the longest is the Leningradsky Prospekt, which runs for 8.6 miles from the center of the city and leads southward; this should not be confused with the Leningrad Chaussée, which becomes the highway, such as it is, to Leningrad.[8]

Historically, Moscow took root on its present site because two small rivers, the Moskva and the Neglinnaya met here, thus facilitating trade. Even today, situated at the junction of the Moskva River and the Moscow-Volga canal, the city owes much to its river traffic; it lies in the middle of the great Russian plain, hundreds of miles from any large body of water or direct outlet to the sea, but it is a flourishing and important port. The actual founder of the city (A.D. 1147) is supposed to have been Prince Yuri Dolgoruki, a feudal lord known as "George of the Long Hands." The community expanded rapidly as a trading post, and became the capital of what was then known as the Principality of Muscovy. Invading Tatars twice burned it to the ground. In 1713 Peter the Great made St. Petersburg (Leningrad today) his capital instead of Moscow, but the Bolsheviks reinstated it soon after the Revolution in 1917. They thought that Leningrad was "foreign," too close to dangerous frontiers, too exposed. Anyway Moscow was the solar plexus of the country because of the growth of industry, particularly cotton textiles.

One of the most interesting—but little-known—sights of Moscow today is the Museum of History and Reconstruction of Moscow, near the center of town. Here the history of the great city is displayed in a series of models, artifacts, and elaborately contrived tableaux. One may see the jawbone of a mammoth unearthed nearby, reproductions of early log cabins, with their intricately dovetailing strips of rough wood, and the types of chain mail worn in the days of Boris Godunov and Ivan the Terrible; then, representing the modern age, such curiosities as the first issue of *Izvestia*, the government newspaper (March 2, 1917), and extraordinarily lifelike papier-mâché models of such events

[8] Serviceable guides to these matters are *Moscow, a Brief Outline* (1967), and the *Moscow Tourist Guide, 1917-1967*. The former is published by the municipality, the latter by the Novosti Press Agency.

as the street fighting in the revolution of 1905, with the Czar's horsemen trampling on the people and the red flag waving behind the barricades. I have mentioned above the inordinate breadth of many contemporary streets. Here are models showing how Gorky Street, as an example, was recently made wider. This important street, once called Tverskaya and renamed for the celebrated writer, is Moscow's Broadway and Fifth Avenue in one. Not a building on its whole length was altered in any way, much less pulled down, but all were physically pushed back a few yards on each side by the ingenious use of rollers set up to operate on underground rails.

Moscow is an astonishingly clean city, and has no smog. This is because no fuel except gas is permitted for either industry or dwellings; the gas is piped in from Saratov, a city 650 miles away on the Volga, or from Central Asia. The comparison to New York, where 85,000 *tons* of soot and other foreign matter fall from the sky every year,[9] is striking. During winter the snow, which may be heavy, is removed from the Moscow streets expeditiously as a rule; no fewer than 75,000 trucks and other vehicles are assigned to perform this service, together with thousands of women with rough, tousled brooms. Muscovites seem to have a passion for covering things up, presumably to keep them clean; parked automobiles are often protected by a large slip cover, like a raincoat, and most passengers on Soviet aircraft encase their hand luggage (if they have any) with a removable fabric jacket. (But there are plenty of unkempt Russians, too.)

This is a silent city. Except for the normal groans and grunts of traffic, now increasing, this is a metropolis (like Tokyo) where things are to be seen, not heard. Drivers are forbidden to use horns except in an emergency, and sounds so familiar in America as shrieking police whistles and the moans of sirens are unusual. No aircraft passes overhead, as I have already mentioned, and children do not yell on the streets. The silence is strange, almost

[9] "Lindsay of New York," by Larry L. King, *Harper's Magazine*, August, 1968.

eerie, because this is very much a city of crowds, where noise
would normally be expected; men and women fairly pullulate
along the streets. As the British essayist V. S. Pritchett wrote
not long ago, "There seems to be no such thing as a Russian
alone."[10] The Russian people are almost as gregarious as Americans, but—another paradox—I felt little spirit of *community*, no
true collective sense. This is certainly not a classless society, in
spite of all pretensions to the contrary. The elite here are con-
spicuously self-segregated, to understate the case. I have been on
a Soviet ship where there were not merely three classes, as on our
transatlantic liners in older days, but actually four: deluxe first,
first, second, third.

But let us return to street scenes. There are signs and slogans on
buildings everywhere, like "Слава кпсс" ("Glory to the Commu-
nist Party of the Soviet Union"), or single words such as "Мир"
("Peace"), "Труд" ("Labor"), and "Май" ("May") during the
May Day celebrations. Portraits of Marx, Lenin, and sometimes
Engels are also to be seen all over the place. One conspicuous
banner on a main road into town says in English, "ATOMS FOR
PEACE."

On a homely level the streets are fascinating, from the beds of
red tulips in the Kremlin gardens to the lines of small prerevolu-
tionary wooden homes, like dolls' houses, with brightly painted
ornate door and window frames, in the suburbs near the airports. Twenty minutes from Moscow I saw a boy placidly fishing
in a trout stream as brown as a walnut table. There are vending
machines on the big streets dispensing kvass and a carbonated
drink called simply *voda* (water), which is water with a sweet
syrup. Old men with gnarled, shaven heads gaze endlessly at
copies of *Pravda* spread out under standing glass frames at street
corners, and crowds of restless young folk line up for lollipops,
beer, and hot dogs—less pink than ours, with two to a roll. Ice
cream cones and other frozen confections are on sale almost
everywhere, even in the marmoreal subway.

[10] *New Statesman*, March 15, 1966.

One minor detail that struck me sharply is that we frequently saw, of all things, pets. Seldom in Moscow before did I ever run into a dog or a cat. Animals did not exist because there was no food for them. But now—to take a single instance—there are no fewer than 150 Great Danes in Moscow, and I even heard that an applicant for a new apartment is asked whether or not he has a pet and how big it is so that extra space may be provided accordingly—if any housing is available at all. Dogs and cats are, however, seldom *seen* on the streets by daylight. This is because most Moscow women have jobs and cannot be at home during the day to walk their animals. The dog is trained to patience and good behavior between dawn and dusk—if you want to believe what I was told.

I never saw a man comb his hair in public, which was an almost universal sight ten years ago, particularly when a person was entering a theater or other place of public assemblage. Conversely, I noticed a considerable number of amputees in the streets, a sight that would have been impossible in the 1950's, because at that time amputees and other cripples had been gathered up and packed off to remote villages in Central Asia, in order that Moscow, Leningrad, and the other large cities in European Russia should not be defaced by them. I never saw a beggar, though—then or now—although some may still be encountered near the churches.

I might add a word superficially on Moscow's illimitable sheer strangeness, its unexpectedness and exoticism, as it confronts the average visiting American. Truly, one is transported here to a different world. I will content myself merely with listing a few small curiosities. There are a good many examples of westernization (e.g., automats, laundromats, and cafeterias), but no Muscovite has ever had a glass of Coca-Cola or played a game of golf. Nobody has ever seen in Moscow a glass-walled bank, bridges with double arches to hold two tiers of traffic, smoke in manholes, a hamburger stand, or a five-and-ten. There are only half a dozen filling stations selling Western-quality gasoline in the whole city, and nobody has ever speculated on a stock exchange or read a gossip column. Then, too, there is the peculiar nature of the look of most people—short, square, squat—hurrying down the streets at dusk almost like a herd of buffaloes.

Several small amenities may be observed. The benches on Gorky Street, where pedestrians may pause and rest, are multi-colored in stripes of the brightest lacquer, like children's toys. In the zoo—not a good zoo—a variety of cubs and small animals of different species are caged together and play amicably, although they will be enemies in later life. The new apartment blocks do not have their entrances in front, but mostly face backward—a traditional rule—so that tenants will be exposed to a minimum of traffic noise; moreover, buildings must stand at least twenty-five feet distant from the nearest street. Women who are guards in the museums, with their seared faces seemingly made of putty, rest their feet on inconspicuous slabs of wood, not on the naked floor. There are, incidentally, few pretty girls (such a con-trast to Warsaw!); many women have hairpieces that do not quite match their hair. Men's trousers are never cuffed—to save material. A wife often has a more important and better-paying job than her husband, but this does not seem to incite jealousy or resentment. I met one woman—an important government official —whose husband is a lathe worker. Conversely, I know one for-eign ambassador whose Muscovite cook is married to a Russian lawyer; they maintain a comfortable dacha in the country.

The current rate of population growth is comparatively low—1.1, exactly the same as that of the United States, which means that it will take sixty-three years to double the population. Life expectancy at birth is seventy, as against seventy-one for the United States. One principal reason why the Soviet birth rate is low is that almost all women work, and do not wish to take time out for bearing children; moreover, they place enormous hope in the future of their children, if any, and calculate that it will be easier to educate and make a better life for small families rather than large ones. The main factor is the housing shortage. By almost universal custom, baby girls are dressed in pink, boys in blue. Only once in Moscow did I see a swaddled child—a familiar sight in older days. Contraceptives are widely used, and abortions are cheap as well as legal. Sexuality is rampant. I asked a middle-aged Russian lady in a restaurant if she thought that there was any girl in the Soviet Union over the age of fourteen who was still a virgin; her answer was to laugh in an outraged manner and slap me with her napkin.

Perhaps strangely, children are not much seen on the streets of Moscow; they go to kindergartens, supported by the state, and, later, to Pioneer camps; during holidays they play in the *dvor*, or backyard, if they have one. The *dvors* of old Moscow are haunts of the grandparents as well, who act as guardians while the parents are at work; in the evening, when the fathers return, they are traditionally the scene of games of dominoes, which, a minor Russian characteristic, are put down with a vigorous, noisy bang. Adolescents are mad about transistor radios; one housewife told me that it was impossible for her to pursue her former avocation of bird watching in the environs of Moscow; there was a youngster with a transistor behind every bush, scaring the birds away. Patterns in courtship have, I was told, changed as well; a boy seeking pleasure from a girl would, in older days, modestly begin by putting his arm around her waist, but now his targets are higher or lower, and impatiently made manifest.

✿

Dope addiction is not a problem in Moscow—at least not yet. Most teen-agers have scarcely even heard of marijuana, much less LSD. Alcohol is, on the contrary, a serious problem among both adults and youngsters, and this is beginning to be admitted. For a long time the authorities denied that alcoholism on a serious scale even existed—who would want to relieve tensions or blot things out by alcohol in a "perfect," classless society? A steep rise in vodka consumption came a year or so ago after introduction of the five-day week; sales of vodka (which is a government monopoly) went up 25 percent, presumably because the worker had nothing much to do in his leisure time except drink himself into a stupor.[11] The city maintains "drunk tanks" to which citizens who are obviously drunk or who have passed out in the streets are taken by the police; they receive a "cure" under medical care in these establishments and are cleaned up and released the next morning, having had to pay a substantial fee. As to juvenile delinquency: this is at a minimum for what seems to be two main reasons. First, school is taken very seriously by most youngsters, homework is arduous, and there is no time for loitering. Second, the Komsomols (Communist youth organization), with 800,000 members, act as a kind of voluntary militia after school hours, helping to police the streets.

Moscow has comparatively little crime, most of it petty. Several times I noticed that friends carefully took off the blades of their windshield wipers and stowed them away in the locked car if we were to be parked for any time; these are a favorite object for theft. We spent one morning listening to a case in a municipal court, and then talking to the judges and assessors, who seemed to be seasoned and temperate men and women. Less than one tenth of one percent of the municipal budget, we heard, goes to the local police—ordinary traffic cops and the like, not the KGB or secret police. Security is, for a normal person, complete; anybody can wander around anywhere at night in perfect safety. A far cry from Central Park in New York!

Moscow has no outright slums, which aids in keeping down crime and delinquency. There are plenty of decrepit houses in neighborhoods less desirable than others, but nobody has what I heard called a "slum mentality," and nobody is a slum dweller from the social point of view.[12] Every kind of snafu has attended public building in the worst areas and elsewhere. One large skyscraper went up recently with sealed windows, but the builders neglected to install air-conditioning, as had been provided for in the plans, and the entire structure, already partly occupied, had to be vacated.

The public services are well maintained, particularly in health

[11] "Russian Weekend," by Denis Blakeley, *The Listener*, June 27, 1968.
[12] Except perhaps a few gypsies.

matters; what is known as "Quick-Help" exists, which gives mobile medical service on the spot, even in people's homes; on the other hand, I heard that to get an ordinary ambulance or even doctor was often difficult—and extremely slow—in an emergency, because the city is stratified by districts and an ailing person cannot be given medical attention out of his or her own area.

The city is convulsed by a fury of building; gaunt yellow cranes are visible almost everywhere, even on the periphery of Red Square; the streets are, however, almost never torn up, or so it seems; nothing exists to compare with the insufferable havoc wrought in New York City by pneumatic drills. One conspicuous new structure is the double skyscraper housing COMECON—the Council for Mutual Economic Assistance among the satellites, which corresponds more or less to the Common Market in Western Europe—on Kalinin Prospekt, or the New Arbat, as it is often called. This is built in the shape of an open book, and is extraordinarily handsome—no touch of nightmare-Gothic here. The building was not, however, built well, or so I was told, and has not yet been able to be used. Nearby is a complex of other purewhite handsome structures, ranging in height from twenty-two stories to twenty-six.[13]

Shopping in Moscow is a complicated story. I have already mentioned GUM, the *valuta* establishments, and the Gastronomes. Altogether there are about six thousand shops—very few indeed by American criteria—including a new supermarket. One astonishing statistic I came across is that between 30-40 percent of all goods sold at retail are imported from foreign countries, mostly from the Soviet bloc. Every single item available in the shops is allocated to its place by the Gosplan, or central planning agency in Moscow; one authority tells me that the Gosplan, which has a comparatively small staff—about two thousand—deals with more than a *million* enterprises all over the Soviet Union, establishing priorities. The resultant copious bureaucratic confusions may be easily imagined. Yet the system works, and the

[13] *Business Week*, October 28, 1967.

concept of planning is basic to all Soviet enterprise. There are, however, innovations from time to time. Considerably more accent is given to the profit motive now than before. Advertising has recently begun to appear in the Moscow newspapers, a startling development, and insurance—on automobiles, personal accident, and property damage—is now being handled by several Moscow agencies. Minor details are that pawnshops flourish and that a car-rental service on our model has been established.

Urban communications are pretty good. Some fourteen *million* passengers are carried every day, 34.4 percent of those by the Metro (subway), 15.4 percent by streetcars, the rest by buses and private cars. At present the average citizen spends from fifty minutes to an hour getting to work; the hope is to reduce this to half an hour by 1980. The Metro, one of the great sights of the city, covers about 150 miles with 85 stations; service stops from 12:30 A.M. till 6, and the fare, for any length of ride, with transfers, is 5 kopecks, a little more than a nickel. I have mentioned the Moscow Metro in the preceding chapter as an equivalent of the Moscow Opera. It is more than that. Its sumptuousness and gleaming marble stations, each different, are a direct expression of Soviet pride and nationalism. In almost every respect—from ventilation to the use of escalators—it makes the New York subway seem inadequate by comparison.

Buses and streetcars cost 3 to 5 kopecks. There are as a rule no ticket collectors on the buses; the passenger operates on the honor system, and deposits his coin in a box, even though this is at the back end of the bus far from the driver. Taxis, of which there are eleven thousand in Moscow, are state-owned, and can be ordered by telephone (something that cannot be said of several great American cities); there are about 150 special phone booths in the streets for summoning taxis. The taxis wear bands of checkers—seemingly a universal symbol. Underpasses exist for pedestrians (as in London and Vienna, but not New York) at several important intersections, and one innovation is the use of portable red railings to extend the width of the sidewalks temporarily if they become crowded.

The telephone service is not too bad, but not as good as ours. For many years no telephone books existed—a major nuisance, prompted by considerations of security—but Moscow finally got a book in 1960. This was, however, exhausted so quickly that many owners of telephones never got a copy, and it has never been reprinted. The result is that it can take days to track a person down. This is, in turn, a reminder of a central, irreversible fact about Moscow—that this is the capital of a country not free. Exciting it may be, industrious, creative, durable, packed with power and force, but it is still a kind of automaton among cities, regulated to the uttermost inch, devoid of the priceless asset of freedom, and governed in the last analysis by the harshest type of fear.

10. MOSCOW II—A FEW
PRIME SIGHTS AND
PROBLEMS

The principal Moscow sights are, as goes without saying, the Kremlin, Red Square, and Lenin's tomb. There is hardly need to describe these in detail since they are universally known. I might add in regard to the Kremlin that the fact is carefully played down that the wall and great cathedrals were built, not by Russian architects, but by Italians imported for the occasion in the fifteenth century; also, that it has a new establishment since my last visit—the magnificent new Palace of Congresses, built of white marble, where the Supreme Soviet sits and where, in between times, theatrical performances are given. The seating capacity is around six thousand, and, when political business is going on, simultaneous translation is provided in twenty-nine different languages. Some aspects of the structure look remarkably like Lincoln Center in New York. Four Americans are, incidentally, buried in the Kremlin wall, among them "Big Bill" Heywood and John Reed.

I heard a nice little story about St. Basil's in Red Square, with its peppermint-stick towers and colored onion domes. Stalin wanted to destroy this irreplaceable monument in order to speed up traffic, and asked a leading Soviet architect, by name Tsetsulin, for his opinion. Mr. Tsetsulin replied that, if this were done, he would commit suicide at the entrance door. Stalin yielded. Mr.

Tsetsulin survived, and, as a matter of fact, recently designed the new Rossiya Hotel, although he is now seventy-five. Nearby facing the square is Lenin's sarcophagus, which draws about nine thousand visitors a day, fifteen thousand on holidays. Stalin was, as everybody knows, similarly pickled and exhibited to the public view next to Lenin, but his body was removed and put in a nearby grave during the Khrushchev process of de-Stalinization. No bust or other memorial marks it. The only other person in the world whose body has been similarly embalmed and exposed to display, like Lenin's, is Georgi Dimitrov, the Bulgarian fire-brand who sensationally defied Field Marshal Hermann Goering during the Reichstag Fire trial in Berlin in the thirties. Dimitrov's body lies in Sofia. The process whereby permanent mummification is achieved is still secret.

A more plebeian Moscow sight is the large swimming pool, built in 1960, which fills a whole city block near the Pushkin Museum almost in the center of town; the largest pool in the world, it holds two thousand swimmers at a time, and is heated the year around so that a haze of steam rises from its waters in winter; yet when I visited it on a perfect day in May, it held no more than half a dozen swimmers. In a similar category is the Central Lenin Stadium, a sports arena said to be the largest in Europe, holding 103,000 people. The race track is worth a visit, too; its totalizer and betting system are identical with those of the bourgeois West. Then, too, one should mention in passing the Ostankino TV tower, said to be the highest structure in the world (1,751 feet), which will have a revolving restaurant on top. Another indispensable sight is the Exhibition of Economic Achievement of the U.S.S.R., formerly known more modestly as the Industrial Exhibition. This covers something like five hundred acres, and takes a lot of seeing; it contains everything from the first Sputnik to devices intended to disprove the assertion that, even if the Russians can cut a notch on the moon, they are incapable of producing a workable corkscrew or pair of scissors. My wife spent a morning here looking over a special display of new consumer goods—furniture, kitchenware, sporting goods, and children's toys, at which the Russians are superb. An enameled saucepan cost 1.60 rubles, a turtle-neck sweater 22.

A list of standard museums and such in Moscow would be endless; there are more than 150 museums, including the Tretyakov, with its icons and other Russian works of art, and the Pushkin, which contains a comparatively small but stunning collection of French impressionist painting. For anybody who does not know the Russian alphabet well the identification of painters can be a puzzling little game; Monet (Моне), Manet (Мане) and Van Gogh (Ван Гог) are easy, but Sisley (Сислей) gave me a moment's pause, and even Picasso (Пикассо) looked cryptic. Another standard sight is, of course, the university on Lenin Hills; it rises thirty-one stories in hideous wedding cake style, covers an area fifteen times greater than that of Columbia University in New York, and has thirty thousand students.

Then, too, there should be a word about the cemeteries. One afternoon we penetrated out of bounds to Peredelkino, the principal writers' village; here a group of Soviet authors of distinction, like the veteran Kornei Chukovsky, live in a kind of pastoral colony. A cop stopped us, since travel is forbidden beyond the rim of Moscow without special permission, which we did not have, but let us go after a mild argument. A woman along the road tried to sell us radishes; there were TV aerials on the humblest hut in a nearby village; nightingales quivered in the bushes; a boy passed by on a motor scooter with a girl and a guitar. We climbed up a muddy hill to reach our objective—the grave of Boris Pasternak. This is marked by a gray stone basrelief in the modern manner, with a simple inscription and the dates 1890-1960. Nearby is the grave of his wife Eugenia; both look over an orchard of cherry and apple trees. The atmosphere was that of unutterable, hard-gained peace. Later in Moscow we searched out the grave of Vladimir Mayakovsky (1893-1930), the poet; a bust—defiant, handsome—on a gray pedestal rising on red stone defines the site. The Russian custom is to leave flowers or other tokens—like eggs and cake—on graves; these are nibbled at by birds, which are supposed to be representatives of the human spirit; by eating the token food they symbolize the will to keep the spirit of the dead alive. Adorning the Mayakovsky bust were a spray of flowers and a red handkerchief. We discovered then that this metropolitan cemetery, near what had once been an Orthodox nunnery famous in folklore, also held the graves of the widows of such important contemporary Soviet leaders as Anastas Mikoyan and the late Kliment Y. Voroshilov. The Mikoyan family has its own plot, with five tombstones so far. We came then to the grave of Stalin's second wife, the mother of the writer Svetlana Alliluyeva, who has recently become so well known all over the world. Her mother rests under a slim white marble post, with a bust on top—calm-looking, with the hair parted in the middle. A translation of the brief inscription reads: "ALLILUYEVA STALINA 1901-31, Member of the Party, from I. V. Stalin." On the grave were fresh flowers and an egg dyed red.

A Word on Government

The Deputy Mayor of Moscow is Anatoly M. Pegov, and his card read (in English): "Secretary Executive Committee of the Moscow City Soviet of Working People's Deputies." The actual Mayor, V. F. Promyslov, was out of town at the time of our visit. Mr. Pegov is not the only deputy or acting mayor; several men have approximately equal status. Another name to conjure with is V. V. Grishin, the head of the *party* organization dealing with the city. In a showdown Grishin, the direct representative of the twelve-man Politburo, which is the ruling body of the entire Soviet Union, and of which he is an alternate member, would outrank anybody on the purely municipal level, although this might not be publicly admitted.

Mr. Pegov is an attractive man, graying, baldish, about fifty, with bright candid blue eyes, thin cheeks, and well-kept hands. His dress was that of any prominent American business executive—dark suit, white shirt, unexceptional tie—and his manner forceful, friendly, and direct. As is typical in Soviet official life, he was punctual to the second, and no interruptions, no tele-

phone calls, no distractions of any kind, occurred during the two
hours we had with him.

We met in an immense room, as big as a tennis court, in the
city hall on Gorky Street, formerly the palace of the governor
general of Moscow. Mr. Pegov comes of the humblest beginnings;
he started adult life as a manual laborer in an automobile factory (his father was a worker, too), rose to be a machinist, took
evening courses at the university, branched out into activity with
the Komsomols, became the representative of a workers' group in
the Likhachev automobile factory, spent eight years as a municipal councilor, kept himself fit by assiduous attention to sport, and
rose at last to his present position—a typical postwar Soviet
career.

The Mayor of Moscow has considerable powers, more than in
an American or West European city, although the national authorities can countermand him. His jurisdiction is wide; he can,
as an example, order the demolition of any building on his own
authority. He runs housing, city planning, sanitation, the hospitals, the markets, schools (but not the university), the maintenance of streets and roadways, the fire department, and the
police if the KGB (secret police) are not involved. He has, however, little to do with labor—such a headache to most American
mayors—because in the U.S.S.R. this is the province of the trade
unions, which have a membership of 86 million in all. "The working people," Mr. Pegov smiled, "are the masters of us all."

The city is divided into seventeen districts, with the Forest
Belt on the periphery as the eighteenth. Citizens elect 1,120
deputies to the Municipal Council every two years; the next elections come late in 1969. Of course, voters have only one party to
choose from. The executive committee numbers twenty-five with
a presidium of eleven. More than a third of the electors are
women, and 48 percent of deputies are workers. Deputies—as in
London and Paris—receive no salaries, and men and women of
all walks of life are represented—teachers, doctors, engineers.

Mr. Pegov offered us mineral water and lemonade, and docu-
mented what he said by a huge wall map, one of the largest I

have ever seen, which lit up in two colors—orange and yellow—to illustrate the points he was making. One color shows the Moscow of today; the other Moscow in 1980. (The lighting system broke down at one interval; two repairmen fixed it in a matter of minutes.) The central core of the whole Moscow structure is, as it is in the Soviet Union itself, organized long-range planning. The radial traffic arteries are to be broadened. The eight railway stations which Moscow has now are to be interconnected. (Conveniently, three of these are already situated on the same square.) The Metro will be expanded from its present network, and trolley-buses are to be abolished, because they tend to strangle traffic. I asked how the streets were kept so clean, and Mr. Pegov, not mentioning the thousands of women who do the sweeping, talked about the large number of machines at his disposal, which douse the streets with water every night. There is no possibility of a water shortage. Delivery of merchandise by trucks takes place mostly at night, although this is not obligatory, to keep the streets freer. Garbage is collected only at night or before working hours in the morning, to eliminate congestion in the streets.

Education is a large preoccupation, and thirty-five to forty new schools, corresponding to our primary or high schools, but not including nurseries or kindergartens, are being built each year. About a million young people go to some sort of school every day, which means that the schools have to operate in two shifts. At present compulsory education is in force for eight years; by 1970 this figure is to rise to ten. Then, too, the municipality has responsibility for a large number of specialized institutions of higher learning, and about 500,000 persons take its correspondence courses. A new art gallery is being planned to expand the Tretyakov, which is swelling at the seams, and a new circus is to be built replacing the present inadequate small building.

Deputy Mayor Pegov called in one of his associates, Mrs. Safronova, the head of the finance department, when we asked about budget figures. She comes of a peasant family, and has worked for the municipality for twenty-two years; her husband is a metal-worker. She had neat dark curly hair, one silver tooth, bright brown button eyes, and a glossy pink manicure; she wore a yellow suit and a small ruby ring. The city's budget is 1.3 billion rubles, more than that of the three Baltic republics, if one may call them that, combined. The equivalent in United States currency at the official rate would be close to a billion and a half dollars.[1] Most, but not all, of this revenue comes from the turnover tax on various enterprises, like the movies, rentals, and profits on the shops. There is no private sector to speak of. Of the city's total revenue, about 45 percent goes into capital expenditure—housing, public buildings, and the like—and about 24.5 percent to hospitals and medical services. Fifteen percent is allotted to education, the maintenance of primary and secondary schools, equipment, and teachers' salaries, and 6 percent to the secondary technical schools.

Other costs include the purchase of electricity, which comes mostly from a grid centered at Kuybyshev, the former Samara, on the Volga. Then, too, there are multitudinous salaries to pay, but these are modest; as an example, Mr. Pegov, near the top, receives only 500 rubles a month. This, at par, would be $5,500 a year; Mayor Lindsay in New York gets $50,000.

Mr. Pegov's last words were a mention of the colossal extent of Russian casualties in World War II, with twenty million dead, and an earnest appeal for good relations with the United States and peace. All Russian officials assert this in almost the same words. If anything categorical can be said about the Soviet Union, it is that no Soviet citizen wants a war.

The Housing Agony

It has always been something of a mystery to me why, in the so-called people's republics, which are regimes theoretically devoted to well-being and the public good, such a prime necessity as housing should always get the short end of the stick. The official

[1] The budget for New York City in 1968-69 was $6 billion.

answer is that the building up of heavy industry, plus the needs of the defense establishment, must be given first priority—guns above butter, et cetera. The real reason is probably that authoritarian governments have, in the past at least, been more concerned about holding on to power than to giving the people an even break. A measure of Khrushchev's quality was that, to a degree, sought to adopt the contrary principle—that fulfilling the needs of the people, so far as this was possible, made the regime stronger, not weaker. He promised meat and milk as well as Sputniks, and his successors have done their best, to a degree, to carry his program on.

Even so, housing is by far the weakest spot in the Moscow picture. There has been a tremendous amount of new building, as previously mentioned in these pages, but not nearly enough. The best proof of this is that 40 percent of the city's families still live in "communal" flats—that is, they do not have an apartment of their own, but share kitchen and toilet facilities with other families, an appalling and shameful situation. Moreover, families which do luckily have their own quarters are stuffed into an intolerably small space; the average is two rooms (not counting kitchen and bath) for three, four, or even more persons. Marriages are compromised by the lack of privacy; divorce is sometimes made impossible because the divorced person can find no new place to live. Sometimes a blanket is hung between the beds of persons just divorced. The Moscow authorities say in defense that about 450,000 citizens have been moved into new apartments since 1960, and that "adequate" housing for all will be assured by 1973-75. The aim is to give every inhabitant 15 square meters (12½ square yards) of living space, about the same as in Western Europe, as against 7.4 at present.

Rents are calculated at a basic rate of 13 kopecks per square meter per month, with a small extra charge for central heat. Water, cold and hot, is free, and, as we know, the total rent may not exceed 6 percent of the income of the head of the family. Most of the new apartment buildings which I saw were made of concrete or cement in prefabricated units. Steel is little used.

Standards of work and maintenance are low; moreover, Moscow is built on marshy ground for the most part, and even the best buildings sag and crack after a year or two. Door handles pull off, hinges break. Any building taller than five stories must have an elevator; the taller buildings are popularly called "American-style," but the official terminology is "International." In general, new apartments are allocated by need; further considerations are a good civic record on the part of the applicant and the circumstances of his present housing. Some persons, like atomic scientists, are heavily favored. The freedom of expression which, on a minor level, exists in Moscow today leads to vociferous argument about rival priorities. If Mrs. S. gets a new apartment ahead of her friend Mrs. F., shouts and screams can be heard practically to the Kremlin.

It is not generally known that co-ops exist in Moscow today for the better-off—also, that there is a considerable amount of private housing, mostly primitive. Private *enterprise* does not exist, but any man who can afford it can buy a house exactly as he can buy a car, although the *land* belongs irrevocably to the state. Lines of these small houses—some made of wood and dilapidated in the extreme—are to be seen near the exterior boulevards. It is often difficult for private owners to command services or get repairs done in a reasonable time.

Jokes

Moscow is not as lighthearted as Warsaw, but political jokes are heard frequently:

"How is Germany divided?" asks a questioner. Reply: "East Germany has the *Communist Manifesto*, West Germany has *Das Kapital*."

"How did the Chinese launch their Sputnik?" Answer: "The entire Chinese nation stood feet on shoulders and *lifted* it into space." Alternative answer: "Seven hundred million Chinese lined up behind it and *blew* it up."

One story has to do with the course of a future Russo-Chinese war. "On the first day the Russians capture a million Chinese. On the second day, five million. On the third, Russia surrenders."

The Theater, Entertainment, and Literary Patterns

Moscow, like all of Russia, is theater-crazy, partly to counterbalance the drabness of environment and daily life, although extravagant private parties sometimes take place. Russians are dramatic people, and passionately like drama. There has always been tremendous vitality in the theater here. People will flock to almost any kind of performance, although their taste is generally good. Moscow has thirty-one actively functioning theaters today, about a hundred movie houses, and several well-known concert halls. Tickets are comparatively cheap (around $5 for the opera), and hard to get on account of the demand. Not through love, money, nor all the good offices of Intourist were we able to hear one Richter concert. There is no such thing as a long "run" in the theater; everything is repertory. The star system is unknown, although a distinguished actor or ballet dancer may be designated as a People's Artist of the Soviet Union, a signal honor, or a People's Artist of the R.S.F.S.R. or other constituent republic of the country.

A kind of off-Broadway theater is flourishing these days, which continually produces new and sometimes exciting talent. The two most interesting performances we witnessed were at theaters of this type—the Taganka, correctly known as the Moscow Drama and Comedy Theater, and the Drama Theater, sometimes called the Malaya Bronnaya. At the Taganka we saw a patriotic pageant called *Ten Days That Shook the World*, which, however, bears little relation to John Reed's book of the same name. Also popular in its repertory are plays by Brecht, Voznezensky, and Peter Weiss. In the lobby stand large portraits of Konstantin Stanislavski, the creator of the modern Russian theater, and of Vsevolod E. Meyerhold, a daring innovator who revolutionized the Soviet stage many years ago, and was liquidated by Stalin as a result. Meyerhold (like Eisenstein in the movies) was a man who really counted, but his methods and techniques have been under wraps for years. The Taganka goes in for a good deal of satire, and has to be careful of the government. It is avant-garde, but not truly experimental. In *Ten Days That Shook the World* a board of footlights is tilted up, making a kind of slanting blurred tent of light as a substitute for a curtain. The acting—as is almost always true in Moscow—is superlative.

The Drama Theater on Malaya Bronnaya is perhaps more serious than the Taganka. Its director is—or was—an adventuresome and talented young man, Anatoly V. Efros. We saw here a highly controversial presentation of *The Three Sisters*, which the orthodox critics severely disapproved of because it dared to take liberties with the classic Chekhov mood. After running in repertory for six months with large success it was abruptly shut down by the Moscow city authorities and the director was fired. Even so, bright youngsters continue to take chances. At a theater called the Satire a play went on for month after month (in repertory) although it had been officially denounced as "pernicious" by the Ministry of Culture.

No one can estimate with much precision what the regime will or will not do about productions it dislikes—it seems to play a cat-and-mouse game. The leading monologist in the country, Arkady Raikin, who has illimitable prestige as well as popularity, can get away with almost anything—even barbed jokes like: "Thanks to Marxism-Leninism, the year consists of four seasons," or: "The party teaches us that water is composed of hydrogen and oxygen."[2]

The standard classic theaters are, for the most part, thought to be old-hat these days—even the MHAT, Moscow Art Theater, founded by Stanislavsky. But we saw there in the uncomfortable old house a production of *The Brothers Karamazov*, which, for sheer magic of realism and quality of acting, equaled anything I have ever seen on a stage. Only three Moscow theaters—an incidental point—predate the revolution: the MHAT, the Bolshoi, and the Maly, a smaller version of the Bolshoi. One theatrical

2 *Newsweek*, January 29, 1968.

experience not to be missed is the Puppet Theater directed with genius by Sergei Obraztsov, People's Artist of the Soviet Union. Moscow also has an interesting gypsy theater, the only one in the world.

As to the Bolshoi, it is still probably the grandest theater on earth, with its six large tiers of boxes, gold-faced, with crimson curtains over white walls, its imposing heavy red-gold curtain, and its sweeping rows of red armchairs, which do not tip up, for the audience. An inconspicuous gutter of chicken wire girdles each majestic circle of boxes, to catch a pair of opera glasses or handbag that might be inadvertently dropped. Light, graceful chandeliers, with seven candles, separate the boxes. One convenience is that the program tells you when the performance will end. In the café each serving of ice cream is individually *weighed* after being scooped out of a container, a clumsy procedure which takes a lot of time and makes for long queues; champagne costs 93 kopecks per glass, about a dollar. The repertory is severely conservative, with experimentation frowned upon, and seems to alternate between old dogs like *Aïda* and *Swan Lake*. But Russian works not often seen outside the Soviet Union, like Tchaikovsky's *Czar's Daughter* and Rimsky-Korsakov's *Sadko*, are available from time to time, and the ballet is still magnificent—if a bit old-fashioned—one of the most thrilling in the world.

TV in Moscow is nothing to shout about (color TV is still experimental), and the current crop of movies is not distinguished except for *War and Peace*, which is quite possibly the greatest movie ever made. We went to a new rendition of *Anna Karenina*, starring Tatyana Samoilova and directed by Alexander Zarkhi, which was disappointing. The plot is twisted to make Anna's husband a sympathetic character, thus emphasizing the contemporary Soviet concern for morality and the sanctity of the home, and the Vronsky was badly cast. But I have seen six or seven Vronskys without encountering one I could believe in. The country's foremost ballet star, Maya Plisetskaya, with her face such a wonderful compound of the sinister and dulcet, makes her debut as a dramatic actress in this film, playing a minor role.

Literature in Moscow is, of course, a whole complex long story in itself. The intellectual climate is lively, within the circumscriptions of dogma, and there are an astonishing number of good bookshops in the city. More than seventy thousand different titles are published in a year, and books are cheap—within the reach of almost everybody. I browsed one morning in a very large, brand-new bookshop on the Kalinin Prospekt; it sells sheet music, post-cards, records, posters, stamps, stationery, as well as books, and is obviously a well-run and thriving establishment. It was pleasant to see what ample attention was given to children's books, and how nicely these are produced. The adult best seller of the day is, I was told, the first book in prose by a poet, Vadim Shefner. I saw little by any contemporary Soviet writer well known in the West, but there are, of course, multitudinous examples of Tolstoi and Maxim Gorky, at prices ranging from 27 to 82 kopecks a volume. A new novel or book of poems will cost under a dollar as a rule. Books are, as is well known, put out in very large editions in Moscow; the first printing of a popular author may run to 500,000 copies, and it may be exhausted the day after publication. This is possible because, as I have just said, books are sensibly made so cheap. Moreover, they are thoroughly read. Books in English were largely technical, like *Corneal Transplantation in Complicated Leucomas* and *Elements of Railway Surveying*. The only translation of an American work of fiction I saw was of Theodore Dreiser's *Sister Carrie*, but I was told that plenty of Arthur Miller was available.

Repercussions of the Daniel-Sinyavsky affair were still being heard vehemently during our stay. This story is too complicated to tell adequately in this space, but the gist is that two Moscow writers, Yuri Daniel and Andrei Sinyavsky, were arrested, tried and sentenced to prison on the charge of transmitting anti-Soviet documentation to the West. Then four other writers, including the poet Alexander Ginzburg, were arrested for activity in support of Daniel-Sinyavsky and for other behavior considered subversive, including editorship of *Phoenix-66*, a widely circulated underground magazine. The central issue became that of differentiating between dissent and treason. Ginzburg received a five-year sentence in a labor camp. Then a courageous young man, Pavel Mikhailovich Litvinov, the grandson of the Soviet Foreign Minister in the 1930's, rose in protest and broke open the whole story by publicizing it in the West, which made an international *cause célèbre*. Litvinov's action was unprecedented in the entire history of the Soviet Union. Never before had an individual openly defied the authorities in this manner. Moreover, he got away with it. If he had acted as he did under Stalin, he would, of course, have been instantly liquidated. Today he walks the streets of Moscow quite free, although he was brutally interrogated by the KGB and lost his job as a teacher.[3]

The importance of this affair is that it confirms several contemporary realities of considerable interest. One is that the youth of the country apparently had little if any knowledge of the Stalin terror thirty years ago. They genuinely—if naïvely—believed that the present regime was based on law and order as it claims, and that constitutional procedures were guaranteed and would be observed. It horrified them that Ginzburg had been held in jail for eleven months before his trial, in clear violation of his constitutional rights, as a recent article in the *New Statesman* points out, and that the trial itself had been viciously unfair—even that Litvinov had been picked up and quizzed by the KGB. A great many people, in other words, have been hoodwinked or brainwashed in recent years into forgetting—or not knowing—that the Soviet Union is still a police state, still absolutist, in spite of its protestation of democratic methods, whenever it wants to be or when circumstances make this, in its view, necessary.

Even so, the intellectual, legal, and moral atmosphere of Moscow is freer than it was a decade ago. The government will not tolerate any "revolt of the intellectuals" or endure being pushed around, but it does not wantonly choose to do much pushing itself except in an emergency, which the Daniel affair was judged to be. In several literary fields there has been a new

[3] Litvinov was, however, subsequently arrested during the crisis over Czechoslovakia in August, 1968.

stringency of control and a tendency to crack down on writers indiscriminately since a party meeting in April, 1968, but also—paradoxically—a bit of loosening up. For instance, after several decades permission has been granted to translate and publish Ernest Hemingway's *For Whom the Bell Tolls*, which was outlawed, despite Hemingway's acceptance and popularity in general, because it revealed the extent of Soviet participation in the Spanish Civil War. Another is that authority has recently been given for the republication in a large edition of the novel *Not by Bread Alone*, by Vladimir Dudintsev, which has been officially considered to be a book "vilifying" the Soviet Union since its first appearance in 1956. At the moment of writing it is unknown whether Aleksandr I. Solzhenitsyn's superb *The First Circle* will be allowed to appear in Russia or not. All this being said, it must be repeated that the regime, if challenged, will continue to insist on retaining all its inhibitory prerogatives in the realm of culture and the intellect.

✿

So much for Moscow. And now we proceed to explore cities in a totally different world, that of the Middle East.

11. THIS YEAR IN
JERUSALEM

The principal preliminary point to make is that Jerusalem is under one roof at last as a result of the Israeli victory in the Six-Day War of June, 1967. The city had been divided between two countries, Israel and Jordan, for almost twenty years, as almost everybody knows—sawed apart into Jewish and Arab sectors. An entire generation of Jews in Israel grew up unable to visit—or even see—its own most sacred relics in the Old City, not even the Wailing Wall, because the Jordanians would not allow them entry, and on the other side Palestinian Arabs could not return to their homes, which in some cases were only a few yards away. Now the city has become whole again.

Jerusalem, with its luminous gold-and-silver sheen, is one of the most beautiful cities in the world—also the oldest, with the possible exception of Damascus. The line goes all the way back to King David, who made it his capital in the tenth century B.C. Here a visitor can walk three thousand years in a minute. But Jerusalem is the capital of a pioneer country as well—which makes a central paradox. Here, shrouded in elusive myth, lie monuments to some of the most revered traditions of mankind, but Jerusalem today is also the fulcrum of sharp political turmoil inextricably involved with the future of the Middle East.

Then, too, Jerusalem is the sacred house of three of the master religions of the world—Christianity with its Holy Sepulcher, Islam represented by the Dome of the Rock, and Judaism with its hallowed ground sacrosanct to Jews. Moreover, several cherished edifices lie within a few yards of each other and contain an intricately mixed and inseparable symbolism. The Church of St. Anne, where Christians worship, has a Jewish origin, became a school under the Arab warrior Saladin, and centuries later passed into the possession of the White Fathers. The Mosque of Aqsa, built on the site of Solomon's Temple, became a Christian basilica, and then marked the spot where the Prophet Mohammed's winged horse was tethered before the flight to heaven—as such, it is the holiest shrine in Islam after Mecca and Medina. The Dome of the Rock, with its cap of gold, was built on the site of Abraham's rock in the Jewish Temple, which became the Haram esh-Sherif, one of the most sacrosanct monuments of Islam. But part of its outer wall is the Jewish Wailing Wall.

The Old City, still walled, still holding the Temple area, is penetrated now by seven gates—Zion Gate, Dung Gate, Damascus Gate, Jaffa Gate, and so on. The Via Dolorosa, with its stations of the cross, where Jesus walked his last bruised steps, lies immediately north of the Temple and is a kind of boundary to the Moslem quarter of the Old City, which contains Christian, Armenian, and Jewish quarters as well. Near are the Mount of Olives, Gethsemane, Mount Zion, Calvary, and the Coenaculum, or Hall of the Last Supper. None of these sites is, of course, archaeologically provable to the last degree. The present-day traveler can do no more than point his finger through mist. But this doesn't matter; what counts is that no city in the world, even Rome, has such a close-knit concentration of ancient and revered monuments in so small a space.

What counts, too, is that Jerusalem is more than a mere city to its Israeli residents—it has a special inner meaning, a spiritual significance. Representing the heart and essence of Jewry, it seems to be projected out of the Book itself. Here past, present, and future merge, as thousands of Jews all over the world intone "Next Year in Jerusalem" at Passover. As one recent observer put it, Jerusalem is "a state of mind rather than a place."[1] Some Jews

[1] *Newsweek*, May 22, 1967.

arrive here on a temporary visit and never leave, because of the spiritual rejuvenation and affinity they come to feel—they develop a compulsion to remain, and call themselves "guardians" of the city, although they are not necessarily either Zionists or religiously observant Jews. To many Arabs, too, Jerusalem has a compulsive bond and holiness.

I asked Teddy Kollek, the Mayor of the city, how he defined "Jerusalem." He gave us five different meanings. Jerusalem is (a) the capital of the state of Israel; (b) the epicenter of the entire Jewish world; (c) an important center for both Moslems and Christians; (d) an international metropolis like Rome or ancient Athens; (e) a place for which the Israelis have several times proved that they were prepared to shed blood.

I asked Abba Eban, the country's Foreign Minister, the same question, and he replied that there were two Jerusalems. First, that of the spirit, or legend, of the imagination, which embodies a strong religious streak and appeals to people everywhere in the world. Second, the concrete Jerusalem of the present, which, despite the sentiment just noted, is abandoned by everybody as soon as Israel gets into trouble. Mr. Eban gave us an ironic, polished smile.

The Israelis won several areas as well as the Jordanian sector of Jerusalem as a result of their swift, electrifying victory in the Six-Day War—the West Bank (of the River Jordan), the Golan Heights on the edge of Syria, and the Sinai Peninsula including the Gaza Strip. These territories may—*may*—be returned or handed over to some other administration in time, but there will be no evacuation without security; the basic political issue in Israel today is whether to annex them or not. We shall come across this issue several times in the pages that follow. But Jerusalem itself is not an issue, as the Israelis see it. East or Jordanian Jerusalem, the Old City, is not negotiable, though it may conceivably be given a different status in the future. But no part of Jerusalem will ever be voluntarily given up. It will remain whole and undivided, if only because Jerusalem *is* Israel.

A Few First Things First

Jerusalem lies 2,500 feet high in a cup of shallow hills, and from several vantage points, Mount Scopus as an example, it gives forth an incomparable radiance. The prevailing colors are cypress green and amber. At night the Old City is illuminated, and its walls, seen from the terrace of the King David Hotel and elsewhere, glow like molten chalk. When Kaiser Wilhelm of Germany visited Jerusalem in 1899, he was reputed to have said, as he stood imperiously near the Jaffa Gate, "This country will go to those who will give it shade."[2] Indeed, the Israelis have planted trees and shrubbery all over the place in the last twenty years, but more green would do no harm. One thing that occurred to me was that Italian masters of the Renaissance, who unendingly painted scenes of the early Christian era in Jerusalem, never had much idea of what it looked like—they made landscapes dotted with towers, columned courtyards, and trees like green muffins. They even had the light wrong, making the city appear to resemble a bland, mildly verdant town in Tuscany. But Jerusalem is whiter, drier, closer to the desert, more gaunt, with a blasting fire of sunshine.

Another first impression is of mixed-upness. This city is both Ford and camel. (Perhaps I should say donkey; I saw a camel only once.) Still another is that there are very few airplanes overhead; Jerusalem does have a small airport, but major aerial traffic puts in at Lod (Lydda), some thirty miles distant by a corkscrew road. Until recently airplanes were not allowed to fly over Jerusalem, for fear that this might violate the armistice agreements made under the former administration; I met children who had never seen an airplane till the Six-Day War. One more curiosity is that Jerusalem is one of the few capitals in the world not bisected by a river or lying next to water in the shape of a sea or a lake. Its hinterland is pure desert, although irrigation has produced

[2] *Israel,* by David Catarivas, in the Vista series, Viking Press, New York.

greenery on the western side. The Mediterranean is about two hours away by automobile, the River Jordan and the Dead Sea, a lugubrious sight, with its water like hot syrup, about half an hour.

Although Jerusalem is one of the most politically sensitive and important capitals in the world, it is quite small, with a total population of about 270,000. Tel Aviv (389,700) is larger, and so was Haifa (207,500) before the war. Of the 270,000 residents of today's Jerusalem some 200,000 are Jews, living mostly in the New City. There are about 55,000 Moslem Arabs, mostly in the Old City, and perhaps 12,000 Christians, mostly in East Jerusalem. Only a handful of Arabs lived in Israeli Jerusalem before the Six-Day War. As to the Christians, who are mainly Arab by ethnic definition but who have adopted Christianity as a religion, they are subdivided into no fewer than twenty-four different sects —Greek Catholic, Greek Orthodox, Roman Catholic, Maronite, and so on—together with several Protestant communities and representatives of the Eastern churches—Armenians, Gregorians, Copts, and Ethiopians. There are also a few Druses (Moslems who broke away from Islam in the eleventh century, and who still form their own fiercely different and tenacious community) and a scattering of Baha'ists.

The focus of most Christians is, naturally, the Church of the Holy Sepulcher, which has been the seat of prolonged and unrelenting jurisdictional tussles between rival Christian sects for many years. Neither architecturally nor aesthetically is it particularly distinguished. Much of it is held up by scaffolding because fierce schismatic quarrels have steadily impeded restoration and repair. The three principal religious groups involved are the Greek Orthodox, the Armenians, and the Franciscans, who represent Roman Catholic interests. But the Copts, Syrians, and Abyssinians are active in its functioning as well. "In the past," Patrick O'Donovan wrote recently in the London *Observer*, "the disputes over passage and property caused sacred processions to clash like foot patrols, with Franciscan friars and Greek priests bashing one another with a fervid but unedifying zeal."[3] Tensions have, however, tended to relax since an ameliorative visit by Pope Paul VI to Jerusalem in 1964. The Pope consulted amicably with both the Orthodox Patriarch of Constantinople and the Patriarch of Jerusalem, by name Benedictos. Incidentally, the Pope and his entourage of Cardinals were forced, like everybody else, to go through the formality of a passport control and customs check at the Mandelbaum Gate before the Jordanian Arabs, who then held power in the Old City, permitted them to enter.

The population of Israel in its entirety (excluding the occupied territories) is around 2,700,000, so that Jerusalem accounts for almost exactly 10 percent of the whole. There are 2,400,000 Jews, 277,000 Moslem Arabs, 70,000 Christian Arabs, 31,000 Druses, and 12,000 Christian Protestants of other sects. The Jews, it will thus be seen, outnumber the Arabs by roughly nine to one. The Arab population increased from about 500,000 at the time of the Balfour Declaration in 1917 to roughly 1,200,000 at the outbreak of the war in 1948, by which Israeli independence was assured, and then sank. In fact, the fighting in 1948 displaced more than 900,000 Arabs—some fled and were not allowed to return, some fled and did not wish to return. A great many became refugees—for instance, in the Gaza Strip, which at that time was not part of Israel—and have been refugees ever since. Others flocked to the West Bank of the Jordan in Jordanian territory. Some were assimilated in other Arab countries. Then after the Six-Day War in 1967, the Israelis found themselves overnight in possession of the Gaza Strip, the West Bank, and other territories, with their great masses of Arab refugees, all of whom now became *Israeli* charges. So all in all, Israel has today the administrative responsibility for 537,000 Arab refugees on the West Bank, 180,000 in the Gaza Strip, 33,000 in northern Sinai, and 6,400 in the area of the Golan Heights in Syria—roughly 756,400 in all, in addition to its Israeli Arabs at home.

This helps to explain something puzzling—why the Israelis

[3] February 25, 1968. Also see Terence Smith in the *New York Times*, June 2, 1968.

have not proceeded to annex the territories they conquered, except for the eastern (Jordanian) sector of Jerusalem, in 1967. One reason is ethics; another is the expense involved. But the principal reason is quite something else again, namely, that to annex these regions would increase the Moslem Arab population of the country by three-quarters of a million people. The Israelis would no longer outweigh the Arabs by nine to one; the ratio would change to something like two and a half to one, and it would be difficult to maintain that Israel was still a National Home for the *Jewish* people, which is of course a cardinal tenet of Israeli doctrine, a fixed tenet of Israeli faith.

There are, however, powerful figures in Israel's public life who feel that this risk must be faced and taken and that, if only for strategic reasons, annexation must be carried out. Then, too, several suggestions for compromise have been put forward. As I have said above, this whole vexing problem of annexation is the principal political issue in the country. It preoccupies almost every citizen.

A few more statistics, cumbersome as they may be, are essential. Israel, like the United States, is the product of natural growth plus immigration. There were only about 24,000 Jews in the country in the 1880's, about 85,000 when World War I broke out in 1914. At that time, Israel, known as Palestine, was part of Turkey —the old Ottoman Empire. The British, fighting the Turks, sought to liberate it as a natural corollary of their war aims, and did so; General Allenby took Jerusalem with regular British forces marching from Egypt, and, as all of us know, Colonel T. E. Lawrence fanned the spark of Arab nationalism and revolt among *Arabs* living in the Turkish domains. Then, too, a Jewish Legion, composed of volunteer Jews from all over the world, which became a powerful military force, fought with Allenby.

In the meantime, Zionism became a vital constituent in this picture. Zionism began—it might be said—with Moses. Even if Moses himself did not reach the Promised Land, he first emphasized the concrete political actuality of the Jewish desire to possess geographical borders, to have for themselves that most essential constituent of nationalism—a homeland. Modern Zionism began in the 1890's with Theodor Herzl, a black-bearded Viennese Jew of unquenchable conviction. He sought to save Jews from the pogroms of Russia and Poland, and to found in the Holy Land—in his view the only place possible—a National Home for the Jewish People. Herzl set up the World Zionist Organization, and the first Zionist National Congress was held in Switzerland in 1897.

We move on to the time of World War I. On November 2, 1917, came the Balfour Declaration, which remains Zionist scripture to this day. The British promulgated it in London partly for humanitarian reasons, partly to draw support from Jewish communities all over the world, in particular the United States. The major actor on the Jewish side was Dr. Chaim Weizmann, a celebrated chemist and Zionist leader resident in England. One of the great men of modern times, Weizmann was a remarkably skillful and subtle-minded statesman, who was destined to become the first President of Israel decades later. He was an uncommonly sophisticated as well as able person.

The Balfour Declaration is still such a seminal document that the text had better be given in full:

> His Majesty's Government view with favor the establishment in Palestine of a National Home for the Jewish people, and will use their best endeavors to facilitate the achievement of this object, it being clearly understood that nothing shall be done which may prejudice the civil and religious rights of existing non-Jewish communities in Palestine[4] or the rights and political status enjoyed by Jews in any other country.

So the Zionist experiment in Palestine was set up, under British administration. The country became a British mandate under the old League of Nations in 1922. British policy blew hot and cold. The British had an obligation under the Balfour Declaration to assist in the creation of the Jewish National Home, but they

[4] But there is no mention of non-Jewish, i.e., Arab, *political* or *national* rights in this remarkably ambiguous document. The Arab argument is that "the establishment *in* Palestine *of* a National Home" is not the same as "the establishment *of* Palestine *as* a National Home," which is what is evolved.

did not want to affront the Arabs. They lopped off Trans-Jordan, originally part of the mandate, and created this as an independent buffer state. Their immigration policy was sinuous, although the future of the National Home depended on this. In the late 1930's they categorically stopped all further Jewish immigration into the country, as a sop to the Arabs. As a result came a fierce explosion of anti-British terrorism by the outraged Jews. I met, in 1968, at least two men who had been outright terrorists—members of either the Irgun or Stern Gang. One had helped blow up the Hotel King David, the other had assisted in the assassination of Count Bernadotte, the Swedish negotiator who was sent to Palestine by the UN to work out some kind of solution of the problem. Both these terrorists, I found, are now Israeli citizens of the utmost respectability. One works as an editor in Tel Aviv, and the other is a prosperous businessman in Jerusalem. I could not have had a more amiable time in the hours I spent with them.

In any case, about 450,000 Jewish immigrants entered the country during the British administration between 1919 and 1948.[5] Then came Israeli independence, and the gates opened. By the end of 1951 almost 700,000 Jews had come in. A law was passed by the Israeli government, now operating on its own, called the Law of the Return, giving Jews anywhere in the world the right of automatic entry into the land. The figures have gone up and down in recent years. Immigration between 1960 and 1964 averaged about 50,000 a year, but in 1965 it declined to 30,736, in 1966 to a little over 15,000. The 1967 figure was 22,000.

The immigrants are of every class and variety, and in recent years have included large numbers of Oriental Jews—incomers from Iraq, Yemen, and North Africa. The word "immigration" is taboo, and nobody is ever called an "immigrant," since all are technically citizens before they arrive; the term used for "immigration" is *"aliya,"* or "ascent," and the individual Jew, no matter how miserable his circumstances, is known as an *"oleh,"* or

[5] *Facts About Israel, 1968*, an Israeli Government publication, p. 61.

ascender. The incomers are carefully processed while en route, and given housing on arrival—there will be some groceries waiting in the larder, camp cots, and a bit of pocket money. But the impoverished, illiterate Oriental Jews constitute a serious problem in integration. There are two Jerusalems now—Jewish and Arab—and nobody wants a third in the form of an Oriental ghetto.

A Jew actually born in Israel (or Palestine under the mandate) is known as a *sabra*. A conspicuous sabra is General Dayan, the Minister of Defense. Other high members of the government show, as is inevitable, a variety of national origins. Prime Minister Levi Eshkol was born in Russia, as was Zalman Shazar, the President of the state; David Ben-Gurion in Poland, Abba Eban in South Africa, and Mayor Kollek in Vienna. A former Governor of Jerusalem (and its first Jewish governor in two thousand years), Dov Joseph, was born in Canada, and Michael Comay, formerly the Israeli Ambassador to the UN, in South Africa.

A good many journalists come from the United States, like the late Gershon Agron, the founder and editor for many years of the Palestine *Post* (now the Jerusalem *Post*) and a subsequent Mayor of Jerusalem, who was born in Philadelphia. Then, too, many prominent Israelis today were Oxford or Cambridge dons, like Walter Eytan, the Ambassador to France, and one of the creators of the Foreign Office. In general, the governmental elite of today is of Polish, Russian, or German origin—men who arrived as children forty or fifty years ago, fought in the Jewish Legion if they were old enough, and worked their way up in the kibbutzim and other agricultural settlements. The Israeli equivalent to our WASP class comes predominantly from Central or Eastern Europe.

A Bit More Background

The origin of the name "Jerusalem" is obscure; apparently it was first called by its last two syllables, "Salem," and then "Urusalem," which means "City of Peace." Clearly "Salem" is related to "*Shalom*" and "*Salaam*," the Hebrew and Arabic words for "peace." In Hebrew the correct name is Yerushalayim, in Latin and Greek Hierosolyma, and in Arabic El Kuds, which means "the Holy."[6] Its history goes far back, but there is little detailed record before 2500 B.C., when it was the small mountain capital of a Semitic people ruled by the Egyptian pharaohs. And certainly the term "City of Peace" can be given an ironic connotation, considering that Jerusalem has been variously conquered and ruled over by Canaanites, Egyptians, Assyrians, Babylonians, Persians, Greeks, Romans, Byzantines, Saracens, Arabs, the Crusaders, Persians, Ottoman Turks (for exactly four hundred years from 1517 to 1917), and the British. Yet it has never lost its earliest Jewish associations.

Beyond reasonable doubt it was David who made it the capital of the Kingdom of Judah in about 1000 B.C.; the previous capital had been Hebron, which is still a revered and important city. David chose Jerusalem because it held a better central position, dominated the watershed between the Mediterranean and the Dead Sea, controlled the route between desert and ocean, and had not been too closely associated with other tribes. David named it "Zion," but nobody knows exactly why; one theory is that "Zion" derives from the Hebrew word for "desert," another is that it originally meant a signpost or marker. The use of "Salem" and eventually "Jerusalem" came later.

Here, in the majestic procession of years which became decades which became centuries, Solomon built his First Temple on Mount Moriah; here Nebuchadnezzar, King of Babylon, came and conquered in 598 B.C. Every schoolchild has a rough idea at least of some of the prodigious events that followed, like the victory of Cyrus the Persian over the Babylonians, the building of the Second Temple (around 520 B.C.), the conquests of Alexander the Great, the revolt of the Maccabees, the beginnings of

[6] From the Baedeker guide, *Palestine and Syria*, published in 1912, hard to come by these days, and very useful. Also see "Jerusalem, City of Dissension or Peace?" by Haim Darin-Drabkin, *New Outlook*, Tel Aviv, January, 1918.

Roman rule, the reign of Herod, and the crucifixion of Jesus Christ at Golgotha. The Jews made a fierce insurrection against Roman rule half a century later, and in A.D. 70 the Emperor Titus reconquered the city and burned it to the ground.

But the Jews were never conquered, never altogether extinguished, even in Jerusalem. We do not need to go into the vicissitudes they managed to survive all over the world in the centuries that followed—wanderers without a land. In the 1930's, some 2,900 years after David, came Adolf Hitler, and six *million* Jews in Western, Central and Eastern Europe were murdered before his evil course was run—the worst catastrophe in the history of Jewry, one of the most appalling tragedies ever to afflict mankind.

The British decided to give up the Palestine mandate in 1947. They were sick and tired of its spiny problems, and eager as well to divest themselves of some of their imperial and colonial commitments. The UN worked out a partition plan dividing the country between Jews and Arabs with Jerusalem internationalized. So independence came to the country once called Canaan in 1948—after eighteen hundred years—and it was renamed Israel. At its narrowest point the new state was only ten miles wide, and the total Jewish population was only about 650,000, less than that of such a city as New Orleans today. It had no single connection by road with the outside world, except through forbidden Arab territory.[7] Instantly its Moslem neighbors declared war on the new republic, determined to kill it off at birth. But it survived, after severe fighting, and so the path hacked out by Moses and Abraham almost four thousand years before reached its goal at last.

A Few Dates as Mileposts

For reasons of clarity and recapitulation, it might be noted that Israel has fought three wars so far:

1. 1948. The new country was, as just mentioned, attacked at

[7] Catarivas, *op. cit.*, p. 126.

once by the regular armies of Egypt, Jordan, Iraq, Syria, and the Lebanon. Although Israel was a stripling, it won in a walk.

2. 1956. Israel participated with Britain and France in the ill-fated attack on Suez.

3. 1967. The Six-Day War in June, which Israel won with a snap of the finger, against Egypt, Syria, Jordan, Iraq, Kuwait, and Algeria. A cease-fire came quickly, but there has been no peace treaty and technically a state of war still exists between Israel and the surrounding Arab states.

For Israel, the smashing victory in 1967 was a tremendous boost. Abba Eban said proudly, "We can breathe with two lungs again," and access to the holy places in Jerusalem became "universal and complete."[8] A dramatic moment came when Teddy Kollek, the Mayor of Jerusalem, stood side by side with General Dayan on what had been the frontier between Israel and Jordan near the entrance to the Old City, and said, "We've got this far, we have to go ahead"—and the bulldozers went to work. What he meant was that the obstructions between the New City and the Old had to be removed. Presently activity began to unify the police of the two administrations, the water supply, the postal system, the telephone lines, and so forth. Plans were advanced to replace no-man's land, still mined, with a public garden. There were no untoward incidents or violence of any kind (these began later), and the Israeli Arabs, who had not stirred during the six-day fighting, remained passive. They did not in any way seek to make a Fifth Column to help their Moslem brethren. "What an occupation!" one observer exclaimed a few months later. There on the streets of the Old City were lines of Arabs, sitting outside the coffee houses and listening placidly to Cairo, Damascus, and Amman on the radio, all of which were pouring out blatant anti-Israeli propaganda. But they were not forbidden to listen or otherwise disturbed, because the Kollek-Dayan policy was not to interfere, out of the hope of placating the Arabs insofar as this was possible.

[8] *Time,* August 4, 1967.

✿

But this does not end the story. As a writer put it recently in the *Times Literary Supplement* of London, "If the Arabs found out that Israel could not be defeated in war, Israel learned that even a victorious campaign could not force the peace."[9] There came ugly episodes of terrorism by Arabs in Jerusalem and elsewhere, and raids and counterraids became endemic along the new frontiers.

Profile of the Golden City

This is a sunny city in several senses of the term. Children look marvelously healthy, and people smile. The atmosphere is full of energy and bounce, and the climate is glorious most of the year. Almost everybody whom one sees in the New City is Jewish, but many do not *look* Jewish, as has been pointed out several times by good observers. Recently I saw photographs of the last Miss Israel—a glorious blonde, who also happens to be a Catholic Arab from Haifa. Another pleasure is that practically everybody speaks English, the universal second language after Hebrew; and it is a distinct and welcome satisfaction to be, for a change, in a foreign country where Americans are liked. Israel could not, of course, survive without financial aid from Jewish and other sources in the United States, but I think that the Israelis would like Americans anyway. Several characteristics—like curiosity, openness, pleasure in gregarious encounters, and practicality—are much the same in both countries.

Jerusalem does, however, give out a decided Oriental flavor even in its western quarters. Ben Yehuda Street and Jaffa Road, two of the main thoroughfares, look more like streets in Baghdad than in New York. There are stalls selling *gazoz*, a raspberry-flavored carbonated water, and open sheds displaying trays of *felafel*, a kind of vegetarian meatball made of chick-peas and

[9] June 4, 1968.

peppers. Arabs wearing the kaffiyeh, or white headdress bound with black rope, are to be encountered frequently, and at regular intervals five times a day the muezzin, or call for prayer, toward Mecca, may be heard from a nearby mosque. This is sounded by a radio loudspeaker nowadays in almost all the mosques, not by an individual priest or imam.

The New City has some thrilling examples of modern architecture, like the National Museum (built in the form of a series of flat boxes and full of esoteric treasures such as the bridal dress of Yemenese virgins), the glassbound Knesset (parliament), and the Hadassah Hospital with its stained-glass windows by Marc Chagall, which have been called "the jewels in the crown of Jerusalem." Important buildings are, as a rule, put up on hilltops, so that they seem to be part of the incomparably lovely stone, and are effectively illuminated at night. The zoo is interesting, because it makes a specialty of containing every animal mentioned in the Bible, about seven hundred species in all.

In the Old City many streets are as narrow as stalks, and they twist and climb; some are covered, and some look like tunnels. There are all manner of steep passages and declivities. We walked one afternoon up Suq El-Bazaar Road, with its neat sign, inherited from the British, "RIGHT TURN PRIORITY FOR PEDESTRIANS," and its network of shallow roofed steps. Here is where Herzl first saw Jerusalem. I noticed Hebrew and Arab newspapers sold from the same stand; American cigarettes at 80 cents a pack; a Diners' Club card posted in the doorway of a caravanserai for mule drivers and their animals; butcher shops equipped with refrigerators; stalls selling a favorite tidbit, round rolls encrusted with sesame seeds and served with a hard-boiled egg; neat occidental posters advertising Ponds Almond Cream atop heaps of dried figs, miniature apricots, almonds from the other side of the Jordan (the trucks came through even during the war), mysterious-looking herbs from India, walnuts, vine leaves, and bright orange-colored lentils.

The foremost of the ancient sights is, to my mind, the Mount

of Olives, and after this the Wailing Wall. A large Jewish cemetery, the most sacrosanct in Zion, covered part of the Mount of
Olives; the Arabs destroyed this, and used the tombstones for
paving blocks on the roads nearby. This the Jews will never for-
give, if only because this is the place where they believe that the
resurrection will occur. The Wailing Wall is larger, heavier, more
massive than I had remembered, but short in length. Built as part
of the wall surrounding the Temple area within the Old City, it
is this relic, more than anything else, that tells Jews that they are
veritably living in the City of God. No site in the world, the Jews
say, has absorbed so much agony; no monument has ever given a
people such collective strength. "Millions of Arabs surround us,
but we have no fear," is a watchword enhanced by the Wall.
The square stones near the top are smaller than those below;
they were put there by Sir Moses Montefiore, the nineteenth-
century Anglo-Jewish philanthropist, to keep the edge from crumbling. Filaments of barbed wire, a surviving element of the
British occupation, keep Arab ragamuffins from tossing rocks
down on the Jews praying below. More than a hundred thousand
Jews, half the population of the city, visited the Wall in one day
on the first anniversary of victory in the June war. We watched
some worshipers, while gangs of children played about, and saw
that letters, messages, supplications, had been stuffed into in-
terstices in the blocks of stone, next to projecting tufts of green
moss or grass.

As a curiosity I quote the following passage from Baedeker's
Palestine and Syria, already cited, and published in 1912:

> Descending this lane [Transverse Lane] for four minutes and
> keeping to the left, we reach the Wailing Place of the Jews
> (Kauthal Ma'Arbe), situated beyond the miserable dwellings of
> the moghrebins (Moslems from the N.W. of Africa). The cele-
> brated Wall . . . is 52 yards in length. . . . The nine lowest
> courses of stone consist of huge blocks, only some of which,
> however, are drafted; among these is one (on the N.) 16½ feet
> long and 13 feet wide. It is probable that the Jews as early as
> the middle ages were in the habit of repairing hither to bewail

the downfall of Jerusalem. A touching scene is presented by the figures leaning against the weather-beaten wall, kissing the stones and weeping. The men often sit here for hours, reading their well-thumbed Hebrew prayer-books. The Spanish Jews, whose appearance and bearing are often refined and independent, present a pleasing contrast to their brethren of Poland.

On Friday, toward evening, the following litany is chanted:

LEADER: For the palace that lies desolate;—
    RESPONSE: We sit in solitude and mourn.
LEADER: For the Temple that is destroyed;—
    RESPONSE: We sit in solitude and mourn.
LEADER: For the walls that are overthrown;—
    RESPONSE: We sit in solitude and mourn.
LEADER: For our majesty that is departed;—
    RESPONSE: We sit in solitude and mourn.
LEADER: For our great men who lie dead;—
    RESPONSE: We sit in solitude and mourn.
LEADER: For the precious stones that are burned;—
    RESPONSE: We sit in solitude and mourn.
LEADER: For our kings, who have despised Him;—
    RESPONSE: We sit in solitude and mourn.

Another antiphony is as follows:

LEADER: We pray Thee, have mercy on Zion;—
    RESPONSE: Gather the children of Jerusalem.
LEADER: Haste, haste, Redeemer of Zion;—
    RESPONSE: Speak to the heart of Jerusalem.
And so on, for six stanzas.

Finally, the Garden of Gethsemane (the word means "oil press") should not be missed among sights, nor Golgotha ("Skull" in Aramaic), nor the Dome of the Rock.

✿

Jerusalem has a pleasantly informal atmosphere. Almost everybody is addressed by the first name. My wife and I were having lunch with General Moshe Dayan and Moshe Pearlman, one of his closest associates, when their chauffeur penetrated into the dining room, tapped Dayan on the shoulder, and said, "Come on, Moshe, time to go." The national costume is a white shirt open at the neck, without a necktie. A necktie salesman would starve in Jerusalem. The word *"Shalom,"* which means "Peace," is a universal salutation, used for both "Hello" and "Good-bye." Good conversation is respected, and almost everyone talks a lot; indeed, it has been written that the national sport is eloquence.[10] Among games, if it may properly be called a game, is the Bible Quiz, a kind of competitive countdown in which contestants are queried on their knowledge of the Bible, and which has a national popularity almost like that of a really big baseball or football game in America.

There are three holidays a week—Friday for the Moslems, Saturday for the Jews, Sunday for the Christians. This can make shopping difficult, and otherwise be a nuisance. For instance, observant Jews have no day for sports or travel, since everything is tight shut on Saturday; even public transport stops. Several modern-minded members of the cabinet have talked at times of installing a five-day week, but they get nowhere. Government departments work a six-day week, Sunday through Friday. They say, "Our forebears worked six days, and so must we; the country needs our labor; we cannot afford letting up by having a second free day." This, be it always remembered, is an austere as well as pioneer community. Tuesday is considered to be a lucky day, because it was third day in Genesis, and it is twice mentioned in the Book that on that day "God saw that it was good"; new enterprises are founded, contracts are signed, and marriages take place, if possible, on Tuesdays.

The growth and development of Jerusalem, to anybody who first saw it, as I did, in the 1920's, is literally astonishing. Then it was little more than a cross between an Arab village and a British encampment. I had a long talk—forty years later in 1968 —with Dov Joseph, the tough-minded lawyer whom I mentioned in passing above. He came here as a boy, enlisted in the Jewish Legion, achieved various distinctions, served in a remark-

10 Catarivas, *op. cit.,* p. 188.

able number of cabinet posts—he has been in turn Minister of Supply, Agriculture, Transport and Communications, Commerce and Industry, Justice, and Development—and became military governor of the city after the British left in 1948. He ran the municipality during the great siege in 1948, and has written an interesting book about it, *The Faithful City*. When Mr. Joseph arrived in Israel in 1919, the country had exactly one industrial plant—an ice factory in Jaffa. There were scarcely any automobiles and the peasants still tilled the soil with the most primitive of instruments. There was no electricity—in the whole country—for seven years after the expulsion of the Turks, and no running water for longer than that. Much of the development that exists today was still undreamed of. The Israeli pioneers followed what may seem to be a peculiar set of priorities, building communities first, then roads, then industry. Cotton was introduced, although everybody said that this would be an impossible crop to grow. Wine culture (first set up by the Rothschilds in the last century) and citrus existed, but had not been fully developed. Israeli melons, the most succulent in the world, were unknown.

✴

Automobile traffic is lively these days, although few individuals are rich enough to own cars. Since a good many drivers are undisciplined, accidents are frequent. Driving from Jerusalem to Tel Aviv one will almost always see the cracked carcass of a car recently demolished—not to be confused with the wrecks of vehicles destroyed in the 1948 war and deliberately left on view as a memorial. The accident rate does not, however, come anywhere near that of the United States in proportion to population; when I told an Israeli friend that there were more than five hundred deaths in America from traffic accidents on a "normal" Fourth of July weekend he could not believe that I was telling the truth. Accidents are not as severely punished as in Russia, for example. "We have a certain sympathy for people with nervosity," one friend told us. Almost all traffic cops are army veterans, and many wear military ribbons, like commissionaires in London. Jerusalem has about 160 "Jewish" taxis, about 250 "Arab" taxis. Both go anywhere in the city, but once, outside the King David Hotel, the doorman told us not to take an Arab driver who was first in the rank, because he would almost certainly not know his way to an inconspicuous Jewish restaurant where we wanted to go. There are also a number of "*sherut*" taxis, which carry a number of passengers along a fixed route, like a miniature bus.

Crime is, if I may mention another municipal problem, rather uncommon; serious crime, like murder, is virtually unknown, and there is very little juvenile delinquency. "This is a working city, only twenty years old in its modern phase—people haven't time to be corrupt," Mayor Kollek said to me. "There were a half-million citizens in the streets last week, celebrating our anniversary, and we did not have a single case of drunkenness. Everybody wants to *make the country*." Some police carry Sten guns, because of the current wave of political terrorism, and stand guard over likely targets. One night we heard a big bang—an Arab bomb had been detonated at a filling station three blocks from our hotel. Prostitution exists, but is not conspicuous; a few girls hang around the all-night taxi stands. All in all, Jerusalem is a very relaxed city from the sociological point of view, in spite of its political vibrancy; for instance, I was told that prisoners in the jails are allowed in certain circumstances to spend weekends at home with their families, and that even marriages are permitted to take place between a prisoner and his loved one outside.

Israel has twenty-four newspapers, thirteen of which appear in Hebrew, the others in English, Arabic, German, Hungarian, Yiddish, French, Polish, Rumanian, and Bulgarian. I do not have to emphasize again that this country—smaller in area than New Jersey—is extraordinarily cosmopolitan. There are about 400 other periodicals, 260 in Hebrew, the rest in 11 different languages.[11]

[11] *Facts About Israel, op. cit.,* pp. 172-173.

The paper most visitors know best is the Jerusalem *Post,* edited by American-born Ted Lurie; this has a local circulation of about 22,000 daily, 32,000 Friday, and sells about 13,000 copies to subscribers abroad in a weekly edition. The biggest circulation belongs to a Hebrew daily, *Ma'ariv,* which, appearing at noon, reaches between 150,000 and 200,000 readers. TV has not yet come to the country,[12] but radio, in the form of Kol Israel (the Voice of Israel) is outstanding; there are about half a million radio sets in the country. Its Arabic service, Dar el Iza's el Israiliya, broadcasts fourteen hours a day; services also exist in Ladino, Maghribi, Yiddish, Persian, and "easy" Hebrew, as well as in the familiar Western languages. Bookstores are, I thought, somewhat scant, nor are they well stocked; shortages in foreign exchange may account for this.

The local theater is on the feeble side, although a government-sponsored pamphlet says that theater attendance per capita is the highest in the world, with more than three million tickets sold last year. There are five repertory theaters, of which the best known is the Habimah, or Israel National Theater. We saw one experimental production at the Khan Theater in Jerusalem, an allegory about Noah—dull in the extreme, I thought. Art and music, I discovered, were somewhat controversial subjects. Several veteran Jerusalemites told me that creation in these fields was, strangely enough, "deplored"—on the ground that all creative energy should go into the service of the state, that the real "creators" were men like Kollek and Dayan. And it was put forward that the bent of the Jewish mind was not primarily "artistic," but in such realms as speculation, abstract thought, exploration, science, and, in general, intellectual rather than aesthetic pursuits. Spinoza, Freud, Einstein, Marx, and, in the present day, the archaeologists who worked on the Dead Sea Scrolls are cited.

---

[12] But many citizens possess TV sets and can tune in to programs in Cairo, Beirut, Damascus, and even Cyprus. Hollis Alpert in the *Saturday Review,* March 9, 1968. Late in 1968—since this passage was first written—Israel's first TV station has been set up.

The principal of these, Professor Yigael Yadin, with whom we had a stimulating encounter, is also a leading general in the army, not an insignificant point. (But then practically everybody of consequence in Israel seems to be, or to have been, an army general.) Conversely, it should be mentioned that a large number of supreme executants in the worlds of art and music—from Artur Rubinstein to Leonard Bernstein—are Jewish. Modigliani was a Jew, and Marcel Proust was quarter-Jewish.

The national revival of Hebrew as a spoken language should also be noted in this connection, as well as the enormous acceleration of interest in the "Book"—the Old Testament. As to Hebrew, its proliferation becomes more marked day by day, although, as stated above, practically everybody except the newer immigrants speaks, reads, and writes English. Hebrew is a difficult language, if only because it has no vowels in its modern written form; some newspapers use what is called "voweled Hebrew," in which the letters of the alphabet carry explanatory marks, rather like kana in Japan. Yiddish, which was once a challenge to Hebrew in the older generation, is dying out except among extremely religious Jews, who consider Hebrew too holy to be used generally, and the movement to latinize Hebrew, conspicuous in the 1920's, has long since fallen apart. Indeed, the Hebrew language in Hebrew form has become an essential ingredient of Israeli nationalism, like the Bible. People make little jokes about Hebrew, though. A young man or woman will say, "Hebrew is the language Mother and Father use until they go to bed, but then they speak German [or Russian, or Polish, or Ukrainian]."

Hebrew today is a long distance away from what it was in the time of Moses; about eight thousand technical words, indispensable in the modern age, have had to be invented. The radio uses three different types of Hebrew—Beginners, Middle, and Regular—in order to reach various levels in the population. Numerous schools, called *ulpanim,* have been set up to teach Hebrew to newcomers, using advanced methods of language instruction;

many older men, together with their wives, also regularly attend
ulpanim courses. I asked a young man at the Foreign Office, who
was born in Canada, and whose English is naturally perfect, what
language he used in small talk with the official, a Chicagoan,
who shares his office. Answer: "Hebrew, of course."

The Israeli Army is probably the most powerful single force in
the country—also the most liberal—closely followed by the kub-
butzim, which are the veritable hearthstone of the nation. Both
are essential constituents of the Jerusalem story, but they are so
well known that I mention them here only in passing. One small
epigram describes the difference between the Soviet Union and
Israel, in reference to the kibbutz dwellers: "In the Soviet Union
the shepherds become professors, in Israel the professors become
shepherds." Another element to stress is the extraordinarily high
quality of Israeli officials, both men and women, engaged in practically all realms of the public service, in particular ambassadors
abroad. From the bottom up, most of the citizenry is singularly
competent, seasoned, alert, able, and intelligent. Perhaps this is
a result of the educational system, plus obligatory service in the
army for men and women alike. Free primary education is compulsory to the age of thirteen, and the army itself is as much a
school as it is a fighting force.

✿

Restaurant food is, in Jerusalem, generally undistinguished,
and night life is unknown. The Hotel King David has, however,
a pleasantly run and comfortable small bar and a grillroom downstairs which, I thought, was the best place to lunch or dine in the
city. The other big hotel, the Intercontinental on the southern
slope of the Mount of Olives, is impressively new, ornate, and
shiny. It stood on Jordanian territory (before the Six-Day War)
and was owned mostly by the Jordanian Government, with the
management in the hands of a subsidiary of Pan-American
World Airways. Now, like a good deal of property in East Jeru-

salem, it has been seized outright by the Israeli Government (under procedures which the Arabs hold to be dubious, if not downright illegal) and is operated by the Custodian of "Abandoned" Property, an official Israeli organization.[13] Some Jews will not, however, enter its portals. The young man from the Foreign Office who was our expert, conscientious, and engaging cicerone for three weeks would not even get near it. This is because it lies close—very close—to the Jewish cemetery on the Mount of Olives, and, according to Jewish religious law, nobody may live within the area of a burial ground. Actually, ten yards separate the hotel from the edge of the burial ground, but a side road built by the Jordanians did traverse a section of the cemetery. This is where, as already mentioned, Jewish graves were bulldozed and tombstones destroyed, and, as a result, many Israelis still boycott the hotel.

Several characteristic dishes of the Jerusalem cuisine are Arab, like *hummos,* a bland paste made of various ingredients, mostly beans or chick-peas. *Machshi* is a pepper stuffed with rice, tomatoes, minced meat, and pine seeds; *tahenna* is a composition of ground sesame seeds, oil, and garlic. Arab bread, which is not called "Arab" by the Jews, is shaped flat, like a pancake; the thing to do is dip it into melted butter. Such typically Jewish foods as a bagel or gefüllte fish are locally unknown. There are half a dozen kinds of notably good local cheese, and dishes like shashlik and kabob are as popular as they are in Damascus—or New York. One local specialty has the nice name "hell"—an herb added to Turkish coffee.[14] Jerusalem restaurants are, an incidental point, theoretically tipless, as in Japan, but tourists have spoiled this and almost everybody accepts tips nowadays.

Kosher is a long subject. Here, in a supposedly rational country and city, we enter arcane areas full of totem and taboo. Jewish dietary laws are, as everybody knows, draconically strict; a central rule is that a kosher restaurant—or home—has to have two

[13] Seth S. King, in the *New York Times,* August 31.
[14] *Everybody's Guide to Israel,* by Joan Comay, a delightfully written as well as useful book.

sets of everything, one for meat products, the other for dairy products, which may never be mixed or taken together at the same meal. Still a different set of utensils is reserved for use at Passover. The rules are applied most strictly in such cosmopolitan institutions as the Hotel King David. One night, slipping into the comfortable bar, with its red-jacketed friendly waiters, I asked for a cup of coffee. No. "Why not?" I asked. "Because we cannot mix milk with meat." I said that *I* had not been having any meat. No matter; meat had been served nearby, and no milk could possibly be allowed to enter the same room. It then occurred to me to order *black* coffee. O.K.

The kosher rules are easy enough to learn. No pork; no butter, cheese, or milk *with* meat; no scaleless fish or seafood, such as oysters, shrimp, lobster, and the like; only animals with a cloven foot and which chew cud, thus excluding pigs, deer, and rabbit. Meat must be slaughtered in a particular way, with the blood drained. Margarine is permitted instead of butter, and a powder made of an essence of vegetable plus white of egg is a tolerable substitute for whipped cream. It was interesting to note that many American Jews visiting Israel, who had probably never had a kosher meal in their lives, voluntarily adopted kosher rules for the duration of their stay, probably to give expression to a new feeling of Jewishness and wanting to belong which a sojourn in Jerusalem often inspires. It should also be mentioned that plenty of good Israeli Jews constantly violate kosher regulations. Pork is available almost everywhere, known as "white steak."

Of Jews and Jews

Jewry is an extraordinarily complicated phenomenon; scholars who have given their whole lives to the subject (so I was assured in Jerusalem) do not know all the answers.

The Jewish community here can be subdivided according to several criteria—between those of European origin as against the Orientals, or between those "observant" and nonobservant. Most of the early Zionists were atheists, and there are a large number of atheists in the leadership today. But this does not mean, a complex point, that they are not "religious" to some degree or other, depending on the definition of religion. David Ben-Gurion, for instance, says that it is impossible to be an Israeli Jew without faith, but he is a particular case—a kind of "atheist believer."

My own feeling is that the revival of religion is the single most interesting and significant change I found in Israel since my first visit many years ago. Of course, fixation on religion has always been an important instrument assisting Israeli nationalism, because it enlarges and reinforces faith. But the Book is vitally revered not merely because of its religious significance but as a basic source of history. In it many Israelis, in their search for identity, discover their own roots. The contemporary craze for archaeology is a closely related phenomenon. Everybody digs.

One local expert in these matters told us that there are four main classifications in Jerusalem—religious Jews, nonreligious, irreligious, and antireligious. Another stratification is by seniority —to be among the first million citizens is a distinction. But almost all Jews here have a sense of pride, of accomplishment; they are top-dog, they are the elite; this is their country, and nobody can push them around. Almost everywhere else in the world, as is only too obvious, there can be a stigma attached to being a Jew —but not in Israel. "Here I am a Jew *proudly*," a young newspaperman recently arrived from Chicago told me with urgent excitement.

What is an "Israeli Jew"? Clearly, a Jew who lives in Israel. But not all Israeli Jews are Zionist, strange as this may seem. The Zionists hold officially that Israel is the home for all Jews everywhere in the world, as is made clear by the Law of the Return and the Law of Citizenship; the country belongs to *world* Jewry, not merely to Israel itself. "I am an Israeli first, a Jew second," is a remark I heard several times, and this stress on secular nationalism further complicates the problem. Interestingly enough, the word "Jew" appears only twice in the Israeli Constitution.

One curiosity, both among believers and nonbelievers, is a widespread belief in miracles in the literal sense. I do not mean miracles in the future (a good many Israeli Jews believe that they will be resurrected, and have absolute belief in the coming of the Messiah), but miracles on the political level in the recent past. I have even heard a Jerusalemite say that the murder of six million Jews by Hitler during the war must have had a "meaning"; his interpretation of this was that it provided the final incentive and push to make Israel a viable, independent state. During the war with the Arab countries in 1948 there came an unprecedented rainfall in the Safad region, which resulted in the flight of ten thousand Arabs at a critical military moment. "God was on our side," is the explanation given. There were "miracles," too, in the Six-Day War nineteen years later; otherwise, how could a nation of two and a half million, fighting one hundred million Arabs, have won so decisively in so short a time? Not only were the parachutists and bomber pilots of General Itzhak Rabin, the chief of staff at the time, and General Dayan responsible; so was the will of God.

☆

Observant (Orthodox) Jews in Jerusalem make a fascinating community, and, although they probably do not number more than a fifth of the population, they are a powerful element in its coloration. Rabbis have substantial force. Generally speaking, the observants are intensely puritanical; they are horrified by such spectacles as mixed bathing on the beaches near Tel Aviv. An Orthodox woman, if married, must not show her hair—so at least I was told—and a married woman may not be alone in the same room as a man unless the door is open. Smoking is, of course, taboo on the Sabbath and holy days; even such a hotel as the King David does not permit smoking in its public rooms or restaurants on religious holidays. Most male observants wear a hat, or even two hats (a skullcap underneath the ordinary hat); for a Jew to go into a synagogue without a hat is equivalent to a

Christian wearing one. I was startled, when Teddy Kollek took me to a session of the Knesset, to see that the President of the Republic, making his inaugural speech, wore a fedora.

Kosher rules are, of course, applied with rigor; cooking is not permitted on the Sabbath (Saturday), but dishes prepared the night before may be kept warm, sometimes by a candle. There has been a long controversy as to whether an Orthodox Jew might or might not use electricity on the Sabbath; it has now been ruled that he may do so, on the ground that, by turning on an electric light switch, he is merely activating a mechanism that already exists, not creating something new, which would be forbidden. Worshipers go to the synagogues for half an hour Friday night, then again for one and a half to two hours or longer on Saturday morning. Grace is said after meals, not before. It is not permitted to travel, work, or even use the telephone on the Sabbath, but, in an emergency, *any*thing may be done to save a life.

Suicide is forbidden; so is cremation, but the use of birth control pills and similar contraceptive devices is allowed. Civil marriage does not exist in Israel; participants in a marriage who cannot or do not wish to have a rabbinical ceremony must go abroad. A favorite place is Cyprus. On the whole, it is probably easier for a Jew to marry a non-Jew than for an observant Jew to marry one nonobservant. Divorce is legal, and has been permitted from the earliest times. Another characteristic of observant Jews is their extreme choosiness, if that is the correct word; as an example, the rabbinate does not recognize the Falasha, or Ethiopian Jews, who are black of skin, to be Jews, nor the Karaites, a very dark variety of Egyptian Jew, few of whom had ever left Egypt until 1948. Finally, some Orthodox Jews are resolutely nonpolitical, so much so that they boycott all political activity, do not vote, and do not necessarily observe legislation passed by the Knesset.

The singularity and zeal of the religious Jews have, at one time or other, aroused considerable resentment in other Jews; an organization exists called the League for Fighting Religious Coercion. Substantial numbers of the Orthodox simply have not brought themselves to accept the fact that Israel is *not* a theo-cratic state. The composition of the Jewish community, as well as rivalries within it, is, in fact, a major problem for the Jerusalem of the future, as Mayor Kollek is well aware, and conflicts be-tween the government and the religious community have already taken place, particularly in regard to the supervision of holy relics and religious monuments.

We spent one afternoon in the Jerusalem district called Mea Shearim, the Orthodox quarter. Here are rows upon rows of synagogues, some of them no larger than a room, and a sign warns visitors, "THE TORAH OBLIGES YOU TO DRESS WITH MODESTY—WE DO NOT TOLERATE PEOPLE PASSING THROUGH OUR STREETS IM-MODESTLY DRESSED." Young men wear their hair in curls on the side of the face; those older appear in fur hats and layers of robes even when it is a hundred degrees in the shade. Buildings are crude. Donkey carts have rickety wheels that, in many cases, do not form perfect circles; they have been banged out of line, and creep and crawl along as best they can. The sight-seer is warned, approaching a synagogue, that women must wear sleeves below their elbows, cover their hair, and have stockings on; slacks are not permitted. An automobile traversing the streets on the Sab-bath may be stoned. There are no sermons in the synagogues, and the congregation never kneels; children may crouch, but actual kneeling is not accepted. Chanting is heard, but no formal music. To most of the communicants in this sheltered enclave, which is really a kind of ghetto, the twentieth century has not come; indeed, the world is still held to be flat by some residents of Mea Shearim. Of course, there is a reason behind all this tenac-ity of custom: the observant Jews have adhered to their legends if only because of incessant persecution over the centuries in almost every country. These are still children of the Diaspora.

✿

Finally, one more type of Jew should be mentioned—the émi-gré *from* Zionism. "Brain drain" has become a serious problem, as more and more skilled young men—the best element in the citizenship, engineers, technicians, graduates in the professions —leave the country because it is too small, too poor, and lacks opportunity.

12. A SHEAF OF ISRAELI
PERSONALITIES

General Moshe (Moses) Dayan is intricate, subtle, and extraordinarily accomplished. He has a small, somewhat twisted face, almost like that of a magical child, and his voice is strained. (But perhaps this was because he had just suffered a severe accident when I saw him—a fall in an archaeological dig which almost cost him his life.) He has the courage of ten lions, but looks somewhat delicate. His modesty is remarkable; when, after the accident, he was being driven to the nearest hospital in an ambulance, he refused to permit the driver to use a siren. Dayan is very much a lone wolf, although he is the hero of the country. Unlike almost all his colleagues in the government, he is not an egghead and tends to despise intellectuals; one friend told me—perhaps not quite seriously—that he had never read a book for fun in his life.

The celebrated eye patch is not a deliberate gesture of ostentation, but is worn out of medical and cosmetic necessity. In the middle of World War II the British undertook a military expedition from Jerusalem against the Vichy French in Lebanon and Syria. Dayan, a youthful officer of enterprise, was given the dangerous task of leading a handful of men to penetrate the frontier, capture a bridgehead on Syrian territory, and hold it until Australian units arrived in force to consolidate it twenty-four hours later. This mission he accomplished. Just as it was terminating he was shot in the eye while using binoculars. The glass splintered, causing multiform injuries to the eye. The socket was irremediably damaged—Dayan almost lost his life as well as the eye—and tiny splinters of glass and shrapnel are still there, impossible to remove. No glass eye could be devised to fit the injured socket, and Dayan wears a patch instead. I heard one man say, "If only we could see with two eyes what he sees with one."

Dayan was not a soldier by profession. He was a farmer, an agricultural specialist, and still loves gardens and has a green thumb. He had, however, been second-in-command to the remarkable British soldier and guerrilla leader, Orde Wingate, a passionate pro-Zionist, in 1937. Before this he had been trained in the Haganah, or Jewish militia. The Haganah was outlawed in 1939, when the British Government in London began to clamp down on Israel in earnest, and Dayan was arrested and sent to prison in Acre. Two years later Dr. Chaim Weizmann persuaded the British to let him—and many others—out of jail on the ground that good fighting men among the Zionists could give valuable assistance to the Allied cause. The Churchill government heeded this advice, and forty *thousand* Jews, who had been arrested in Palestine, were released. Dayan promptly became an intelligence officer, and the Syrian episode followed.

Dayan is sometimes nicknamed "the Arab," not because he looks like one but because, in spite of the hawkishness attributed to him by those who do not know him well, he is probably the most conciliatory of all contemporary Israeli leaders in regard to Palestinian affairs. It was he who set up the policy whereby Arabs living in East Jerusalem, now incorporated into the rest of the city, can listen freely out of doors, even in groups, to broadcasts from Arab capitals. He has taken a stand for excluding *Jewish* lawyers from participating in court cases in Israeli cities with a predominantly Arab population, if the Arabs involved don't want them, and he has advocated the utmost freedom of movement between Israel and Jordan although these countries are still technically at war.[1] Hundreds of men and women cross the

---

[1] "A View of Moshe Dayan Without the Bang," by Uri Avneri, *Haolam Hazeh*, Tel Aviv, reprinted in *Atlas*, June, 1968. Mr. Avneri, who describes himself as a non-Zionist Jew, and who has a party of one—his own—in the Knesset, is one of the brightest minds in Israel, although his enemies call him a prankster and crank.

Allenby Bridge, a principal point on the frontier, day by day, carrying on trade. But let nobody think that Dayan is soft on Jordan. The following colloquy took place not long ago between the General and a German journalist, as published in a recent issue of *Encounter.* I paraphrase it slightly:

> GERMAN: How do you hope to achieve peace with the Arabs?
> DAYAN: By standing firm as iron, wherever we are now standing, until the Arabs care to consent.
> GERMAN: Then it's only King Hussein who is likely to qualify as a partner in negotiations. But is he strong enough to agree to your conditions?
> DAYAN: In this case let them find another king.
> GERMAN: But Jordan as a country may not be strong enough to agree to peace on your conditions.
> DAYAN: In this case let them find themselves another country.

Another example of Dayan's style, also culled from a special section on the Middle East in *Encounter,* is the following. Soon after the Six-Day War Dayan was asked about peace talks with the Jordanian Arabs. His laconic answer was that "we are waiting for them to ring us up."

On annexation, the chief political problem of the present period in Israel—what to do about the 26,000 square miles of territory taken from the surrounding Arab countries as a result of the Six-Day War—Dayan is extremely cagey. This is his own particular province, since the Occupied Territories come under his authority as Minister of Defense. A careful reading of his recent statements on the subject, which are somewhat cryptic, seems to indicate that he does not, in principle, stand for annexation—except for East Jerusalem, which will never be given up, and a strategic post at the bottom of the Sinai Peninsula to ensure access to the Gulf of Aqaba. Annexation would, he appears to think, seriously dilute the Jewish majority in the country. What he wants, as a realist, as a pragmatist, is protection. The Arabs in the neighboring countries are, he feels, almost certain to attack again; therefore such an occupied area as the West Bank must become a buffer, a safety zone, a glacis. But how to achieve this without annexation? Therein lies the heart of the problem. One of Dayan's ideas is to amend the frontier slightly to make it more foolproof from the strategic point of view. His *aim* is simple, and never varies—how best to ensure the viability and durability of the Israeli nation. Basically that is all that he is interested in. But, as circumstances and conditions change, so do his tactics. He said once, "I refuse to be the slave of my own speeches." Above all, he seeks to maintain good relations with the Israeli Arabs (those within the old boundaries of Israel), and believes firmly that this is steadily being accomplished.

General Dayan is fifty-two, and looks years younger. His childhood was spent in Nahalal, a remarkable village between Haifa and Nazareth, near the Sea of Galilee. This, built in a circle, is what is known as a "moshav," or settlement, which differs from a kibbutz in that each family owns its own particular property, although work is shared and much endeavor is collective. Nahalal had once been a malarial swamp, but it became—and still is—one of the most successful as well as picturesque moshavim in the country. Dayan's father, by name Shmuel, emigrated to Israel from Russia, at the age of seventeen, and was a poet and writer as well as moshav resident. "He was not a true farmer, not a fighter, just an idealist," Dayan told me. Clearly the General was strongly influenced by his father, one of whose books, *Pioneers in Israel,* is still well known. "He wanted to farm cooperatively, but live independently," Dayan said. The territory nearby was continually harassed by Arab marauders, and young Moshe lived with a hoe in one hand, a rifle in the other. "I was a marksman at the age of twelve," he mentioned during the course of a long lunch. "At sixteen I joined the Haganah."

Dayan's career after World War II may be recapitulated briefly. He went to law school at night (he had never had the opportunity of finishing secondary school), studied political science at the Hebrew university, engaged for a time in a business enterprise, and went to England to attend the staff college at Camberley. Having risen to the rank of colonel in the Israeli Army, he fought in the 1948 war, became chief of staff of the army

1953-58 and commanded the Sinai campaign in 1956, after being commander in chief in the Jerusalem area for a period. Interested in politics and keenly ambitious, he joined the Mapai or government Party, and became Minister of Agriculture in 1959. He was restless, cryptic, and oblique, and he left the cabinet after a time to join a new party, known as the Rafi Party, founded by Ben-Gurion and others. Meanwhile he wrote a book about his campaigns. Israeli domestic politics are confused and internecine to a degree, and Dayan, sitting in opposition, became a continuing shadowy threat to the Eshkol government. Dayan stood for immediate positive action to ward off the Arab (mostly Egyptian) attack that everybody knew was bound to come, as tension developed early in 1967, but the government equivocated. When Colonel Nasser's aggressive intentions could no longer be doubted or ignored, the pressure of public opinion in Israel forced Eshkol to take Dayan back into the government as Minister of Defense. This was greeted by almost everybody as a signal that war was certain to break out at once, and it did. But Dayan and the military did not provoke the war; it was thrust on them.

In previous years Dayan had a good deal to do with the training, development, and morale of the Israeli armed forces. One rule was that every effort must be made, no matter what the cost, to recover any wounded Israeli soldier and the bodies of the dead. The precision of the Israeli strategy and its deadly swiftness of movement are too well known to mention here. Interestingly enough, American military experts predicted that the Six-Day War would last five days; the British said six, and they were right. The Soviet Ambassador to Israel predicted an Arab victory; he was withdrawn from Israel immediately after the fighting stopped, and has never been heard from since.

Dayan during the fighting visited every front hour by hour by helicopter, something that a minister of defense does not ordinarily do. He had the good fortune of having as his chief of staff General Itzhak Rabin, the extraordinarily able military man who later became Israeli Ambassador to Washington. One of Dayan's most useful qualities was that, with his long experience in the army, he knew closely men down to the level of battalion commander or even lower; he would say, "Chaim will be better at this spot than anybody else," or, "Give Jakub his chance here."

Dayan is not formally religious, but on the conclusion of the war he went at once to the Wailing Wall, slipping into it a piece of paper marked "Peace to All Israel." He works in Tel Aviv, not Jerusalem, and spends a great deal of time commuting. Headquarters of the defense forces are located in Tel Aviv rather than Jerusalem because the former is strategically more secure. Dayan has several children, to whom he is devoted. His attractive wife is the director of one of the main handicraft industries in Israel, which has shops in various towns. He talked a good deal about social patterns during our lunch, and one remark I remember was, "I do not think we have any class distinctions in Jerusalem, but if we do, they are certainly not based on wealth." His last words to me were, "I believe in those who believe."

Mayor of Jerusalem

Theodore Kollek, known universally as Teddy, must have more friends than any person in an equivalent position in the world. A stocky vigorous man, gregarious, wary of eye but with a warming smile, his face is bisected by a strong wedge of nose; he wears his brown hair *en pompadour*. Like Dayan, he looks younger than his years—fifty-seven—although his life has been one of struggle, tough and strenuous. He lives on one of the pleasantest streets in Jerusalem, Rashba Road, in an apartment packed with artifacts and works of art, particularly old glass. Here, the legend is, he sits every night with an inner group of cronies, which more or less determines the course of affairs in Jerusalem. His wife, also a Viennese, was a rabbi's daughter; they have two children. He gets up every morning at 6 A.M., and works a sixteen-hour day. His manner is almost always relaxed, supple, informal, and he has an excellent sense of humor.

Mayor Kollek, when my wife and I called on him in his bare, modest office downtown, wore the standard Israeli uniform, a short-sleeved white shirt open at the neck, gray trousers, and brown scuffers. A pencil and notebook stuck out of his shirt pocket. I noticed, having received a letter from him, that "Mayor of Jerusalem" is now printed on his stationery in three languages, Arabic, Hebrew, and English; a letter I had had from him a year or so before did not have the line in Arabic. But now, as we know, there are some seventy thousand Arabs in Jerusalem for whom Kollek is responsible, as against a handful before the Six-Day War. Like Dayan, one of his closest friends, he is tirelessly attentive to the presence, wants, and needs of the new non-Jewish population—under Israeli hegemony, of course.

Theodore Kollek was born in Vienna in 1911, of a well-to-do family, of partly Moravian descent. He did not seem particularly eager to discuss his early days or the details of his career. "Let's get on with it," he kept repeating in his manner both suave and craggy; what he meant was that he wanted to talk about his job, Jerusalem. However, he did give us a bare outline of his background. Vienna was not a happy city when he grew up. It was the intellectual capital of the Jewish world, in a sense, because of the astonishing concentration of eminent Jews who lived and worked there—Freud, Schnitzler, both Zweigs, Mahler, Reinhardt —but it contained as well a strong, menacing note of anti-Semitism. The peace treaties in 1919, signed when Kollek was eight, made the country a geographical monstrosity—a huge head, Vienna, atop a dwarfed, shrunken body. In those pre-Hitler days the Social Democrats favored *Anschluss* (union) with Germany. The youthful Kollek decided to get out, and emigrated to Palestine in 1934.

Virtually all contemporary Zionists of middle age lived in a kibbutz and bore arms in their earlier years; they were both farmers and soldiers, and Kollek is no exception. He became a member of the Ein Gev kibbutz on the Sea of Galilee in 1934, and worked there for some years. Ein Gev, also known as Sussita because it is horse-shaped,[2] is an archaeologist's paradise, where excavations show numerous "layers of settlement" until the community was destroyed by the Syrians in 732 B.C. Later it was called Hippos, also meaning horse. In Kollek's day it was a lively watering place and the home of a music festival. The Arabs attacked it fiercely in 1948.

But this is to get ahead of our story. Young Kollek, who had immense energy, sought further scope. He served with Zionist youth groups in several European countries between 1938 and 1940, and joined the Political Department of the Jewish Agency in 1942. During most of World War II he maintained liaison with Jewish underground groups in Europe, and his gifts as a tough manipulator became well known. He led a Haganah mission to the U.S.A. in 1947-48, seeking to buy arms, and lived for a time in a remarkable small hotel in Manhattan, the Hotel Fourteen on Sixtieth Street, much favored at the time by political émigrés and conspirators. Kollek also served from 1950 to 1952 in a more official capacity as Minister in Washington, after a period devoted largely to the illegal running in of immigrants. Then, back in Israel, he spent twelve years as director-general of the office of the Prime Minister, David Ben-Gurion. During part of the same period he became chairman of the Government Tourist Corporation and led a nuclear project having to do with the desalinization of water. But what counted was his association with Ben-Gurion. They were as close as fingers. He was B.-G.'s indispensable chief assistant as well as warm friend. He took responsibility in B.-G.'s name for administration decisions that B.-G. did not even know about. Ben-Gurion retired, and a confused period followed of political rough-and-tumble; the Labor Party split, and Kollek, like Dayan, joined a left-wing splinter group, the Rafi. He was put up to run for Mayor in 1965, and won handily. He quipped later, "The trouble with running for Mayor is that you may get elected."[3]

[2] Comay, *op. cit.*, p. 355. *"Sus"* is Hebrew for "horse."
[3] Edmund Wilson in *The New Yorker*, August 19, 1967.

Kollek conceived the idea for the magnificent new National Museum, and helped raise the money for it. He is also largely responsible for the creation of the University of the South, or Desert School, in the Negev near the kibbutz where Ben-Gurion lives today.

His difficulties as Mayor have been numerous. First, he always needs a bigger budget than he can get; as an example, he wants to install an improved street-lighting system which would, however, cost about £1,000,000, or $280,000; the money simply isn't there. One must always keep in mind Israel's austerity. (Kollek's own salary is, incidentally, around $500 a month.) He wants to build research institutions, small specialized foundations, and a new theater. Second, the mayoralty is weighted down with various structures inherited from the British, who did not want their Arab puppet-mayors, like the Nashashibis, to have too much power, and the statutes have not been changed since 1934. "My being here is all a mistake," Kollek intones cheerfully. "I'm a fluke."

He says that he does not intend to run for a second term, and has, or so at least I was told, no ambitions to enter the national government. He would, however, probably like to have the office of mayor raised to cabinet rank. His administration is, in any case, well run, and clean as a clean whistle. There are no Daleys in Jerusalem.

Mayor Kollek told us about some of his problems. "We have double the birth rate of Tel Aviv, and yet we have to absorb more immigrants—more children, more Orientals too, which means more illiterates." The altitude is a factor, too; Jerusalem is high and dwellings need central heating in winter, something that Haifa and Tel Aviv do not have to contend with. Jerusalem has an infinite number of public buildings, which the city is proud to support—but they don't pay taxes. The Jerusalem plan provides for a population of about half a million in forty years. The country has no coal or steel, and industrialization must be limited —"We're very choosy"—to such endeavors as electronics, pharmaceuticals, publishing. Jerusalem has "grown up" a great deal,

Kollek thinks; ten years ago he could not walk down Ben Yehuda Street without meeting half a dozen people he knew; not now. "My big difficulties are with the government of Israel, not with the Arabs." In spite of all the provocations on the Cairo radio, not a single Arab was killed in Jerusalem in the period immediately following the Six-Day War; there were no communal clashes of any kind. "We maintain peace by peaceful means," he declared. But the Arabs are in a difficult position, "because we don't want them in our primary schools; we don't want a melting pot." The Mayor's idea is seemingly to give the Arabs a full chance to develop their own culture, with complete civil rights, but they must remain "a permanent unassimilable minority." He added, "It's not impossible. Look at the Alsatians in France."

Mr. Kollek speaks a little Arabic, and goes out of his way to keep in close touch with the Moslem community in East Jerusalem. He drives his own modest car all over the place. He has given the city a new bounce, an acceleration of pace. That is his chief contribution. He has confirmed Jerusalem's position as *the* capital, in spite of the rise of Tel Aviv and despite the fact that the city's position as the capital has not yet received international juridical sanction.

Mr. Eban, the Foreign Minister

Abba Eban, the Minister of Foreign Affairs, was born in South Africa in 1915, and moved to England as a child. The name "Eban" means "stone" in Hebrew. An intellectual of intellectuals, he studied the Semitic languages in England, and taught Arabic for a time at a military school in Lebanon. His Hebrew, like his Arabic, English, and other languages, is of the purest quality. Mr. Eban and one or two colleagues in England were the fathers of the Israeli Foreign Office; they set up its mechanism as early as 1945, before the end of World War II, long before Israel became independent in 1948.

Before this Eban had held various posts—liaison officer to the Allied Headquarters in Palestine in 1940, and an official of the

Jewish Agency in 1946. After independence he became successively Israeli Ambassador to the United States, Ambassador to the UN, which made him an internationally known figure, Minister without Portfolio, and Minister of Education. Glossy, almost burnished, a tall man, with a mind of the utmost subtlety, Eban speaks swiftly in a voice so low that it is almost a whisper, sprinkles his remarks with arcane jokes and indirections, works a long day, and has (like Dayan) strong ambitions to be Prime Minister. He is a most impressive person—intuitive and a bit aloof as well. He is much more right-wing than Kollek or Dayan.

As Foreign Minister he is directly exposed to more contact with other governments than Dayan, and his line is perhaps more pliable. He talked to us at length about the Arab "mystique"—how, among most Arabs, it is what they *say* that counts, not the facts at issue, because, once they have said something, no matter how far off-beam, they believe it. Eban strongly opposes annexation of the West Bank, if I understood his swift, soft-voiced remarks correctly, because of the familiar thesis that Israel does not want 750,000 more Arabs in the country, which would dilute the National Home.

I asked him if the annexation of Arab Jerusalem by Israel was "negotiable." He replied with an ironic smile, "Of course it is *negotiable*, but we would simply say No in any negotiations."

Mr. Eban's erudition is stunning, and his knowledge of languages legendary. When we left him, I said something to the effect that, even if he faced difficult problems, he must be having a good time as Foreign Minister. He replied in his melodious voice, "Well, it's not been dull, and I daresay it is less boring than being Foreign Minister of Norway." I repeated this remark later to an Israeli friend, who responded, "And Abba said this in perfect Norwegian, I have no doubt."

Prime Minister, Party, Politics

A small joke sometimes heard in Jerusalem is that when the Prime Minister, Mr. Eshkol, is asked whether he wants tea or coffee, his reply is "Half and half." Another is that a colleague remonstrates with him for having failed to keep a political promise, and he answers, "Yes, but I didn't promise to keep the promise."

The Jerusalem political arena is full of artful gymnasts in words and creatures of contrivance, but the Prime Minister is more than that. Levi Eshkol (originally Shkolnik), an old-timer, a classic example of the man rising out of the kibbutz, was born in the Ukraine, near Khrushchev's birthplace, but his family moved to Vilna when he was a child, and he was educated at a Hebrew School in Vilna, the former principal city of Lithuania, in the Baltic area.

Eshkol had enterprise enough to emigrate to Palestine in 1913; he worked on various agricultural settlements, and served as a soldier in the 40th Royal Fusiliers under General Allenby. Branching out into politics after this double indoctrination, he became secretary of the Tel Aviv–Jaffa Workers Council and a rising power in both Mapai, the Labor Party, and the Histradut, the powerful National Federation of Workers. For many years he was a conspicuous member of Haganah, the illegal military formation fighting the British. During World War II he spent a good deal of time in, of all places, Berlin, and also returned to Lithuania briefly, in charge of negotiations to transfer German-Jewish property to Palestine. He became Minister of State Security early in the 1950's, then Minister of Agriculture and of Finance, and finally Prime Minister. He has always been somewhat humble in approach, but shrewd, practical, and a good businessman. Sometimes he has been criticized for his tendency to postpone action. His "wall-to-wall" government has lost much of its former authority, because, still a coalition, it covers so broad a front that it has become almost impossible for it to take decisions.

Short in stature, stoutly built, strong, with a beaming eye, wearing an open white shirt (but with a jacket), the Prime Minister opened our talk by reminiscing about early days in the Ukraine. "The British knew very well that we are an Oriental people. Anybody who comes from the Russia of that day is an Oriental." He described—with a vividness which gave a picture both colorful and precise—the way of life of his village, with its flour mill, crossroads, railway station, houses without shutters, cattle, deportations, pogroms. I was much reminded of the contemporary musical comedy *Fiddler on the Roof*. The Shkolniks were rich, in a manner of speaking. They were forbidden to own land, but (he winked) they did, under an assumed non-Jewish name. When he decided to get out, the way was long and difficult, via Vienna and Trieste, and Palestine was still a Turkish fief. Nine boys accompanied him; eight of the nine still live and work here fifty-five years later. "We had to train [drain] the land," he said, because it was suffused with malaria. "I worked with a short pole, a daily laborer." The creation of primitive Jewish settlements followed. "We studied the Bible, the Talmud, Jewish history; I drank in these words like a thirsty man in the desert. The Bible was in my heart. I was a Zionist before I knew what Zionism was. I knew that if we had to revive the Jewish nation, we had to revive it here."

Mr. Eshkol discussed then the possibilities of negotiations with the Arabs today. "When you go to buy a goat, you don't mention the goat first. We know what it means to drive a bargain. In the Gaza Strip they thought that we had horns, but they found out, no. We have to be certain that we can trust their word. We were promised access to the Wailing Wall and the Mount of Olives in solemn agreements in 1948, but the Arabs did not keep their word. I understand the difficulties and dangers of the position of King Hussein. Had it not been for Nasser we might have arrived at a peaceful condition with him long ago. When the Six-Day War was clearly about to break out, we managed to get in touch with Hussein through a UN intermediary, and told him, 'Don't mix in this business.' Hussein made the mistake of his life not to heed this friendly warning. Nasser told him the next day on the telephone that he had already destroyed the Israeli Air Force—ha!—and that he must enter into this thing to be part of the victory. At the same moment Nasser was appealing to Algeria for help, saying that things looked bad."

✿

Many other persons deserve mention in even the briefest portfolio of Jerusalem political dignitaries, like Pinhas Sapir, formerly Minister of Finance, a vital force in the country, who is now secretary general of the governing Mapai Party, as well as his successor in the Finance Ministry, Ze'ev Sharef, the former Minister of Commerce and Industry. But I must, for reasons of space, restrict myself to two outstanding characters.

Yigal Allon, a leader of the former Achdut Ha'avoda (Socialist Party) and Minister of Labor, now Deputy Prime Minister, an Eshkol man and presumably his candidate for the succession, is forty-nine and is a person exceptionally skilled and spirited. He was born in a kibbutz near Galilee, went to Oxford, made a superb war record in 1948 (he was leader of an underground commando unit in Sinai), and did well as Minister of Labor. He stands for "a pluralistic society of Jews and Arabs living together"[4] and recently announced a carefully thought-out plan for the reorganization of the West Bank. This is to be returned to Jordan *in toto*, but Israel is to be protected by a "security belt" of fortified settlements along the border, through which, however, the Arabs will have a protected "gate" to the sea.

Mrs. Golda Meir, who was Eban's predecessor as Foreign Minister for a decade, has often been called "the best man in the country." Once I heard an American, who knew her well at the United Nations, say that she was "the most rigid polemicist" he ever met. Mrs. Meir, who has now retired from active participation in affairs, was born in Kiev in the Ukraine in 1898, but spent part of her childhood in the United States; she was educated in Milwaukee and worked there as a schoolteacher. For many years she was the secretary general of the Mapai Party, and, as such,

[4] *Time*, June 12, 1968.

was the principal political arbiter in the country. Legends about her are without number. Once she went disguised as a man to conduct difficult and totally secret negotiations with the late King Abdullah of Trans-Jordan.

Each Israeli leader has his own pool of private support, and is reluctant to have anybody else swim in it. Unity comes when unity is needed—at moments of grave political crisis, war, or threatened war. "Disaster always unites us," Mr. Eshkol told me. But at other times the ruling team is far from being united or harmonious. Ben-Gurion does not have a high opinion of Eshkol; Dayan and Allon are bitter rivals for the future leadership; Eban and Dayan have differences. Sharp words are often spoken. Teddy Kollek (a Dayan ally) recently criticized the financial estimates prepared by Pinhas Sapir, the Finance Minister, until a few months ago, on the familiar ground that Jerusalem was not getting its fair share of the budget. Mr. Sapir replied, "To say that I do not give help or cooperation to Jerusalem is both a distortion of the truth and a *burning insult* [*sic*]."

✿

Mapai, the Israel Labor Party, has held power uninterruptedly since the formation of the country, but it has never had a clear majority in the Knesset. The reason for this paradox is that elections (every four years) are held under proportionate representation, a heritage from the Zionist National Congress. Proportionate representation is an unwieldy process at best, and serves to make the whole country a single constituency. Mapai is, in effect, the party of the Jewish National Home; what it stands for is the stability of the nation. But, since it falls just short of having a majority, it has always had to rule by coalition; its present partners are two splinter groups, Rafi (Dayan's party) and the socialist Achdut Ha'avoda. Mapai and its allies have 63 deputies in the present Knesset (120 members); there are eleven parties in all. The most interesting opposition party is the Mapam, or

United Workers Party, with nine seats. This, which at one time included the Achdut Ha'avoda, Allon's party, is strongly to the left and urgently wants an accommodation with the Arabs. There are three different religious parties, which take as a rule a decided right-wing position. The Knesset also contains five Arab deputies and four Communists, divided into two groups. Three Communists are Jewish Israelis, one is Arab. Debates take place in Hebrew (not English), with Arabic provided in simultaneous translation; Arab members may speak in Arabic if they choose.

The Old Lion

Perhaps my subtitle here is a misnomer; the actual name "Ben-Gurion" means "son of a lion." Originally B.-G.'s name was David Green, and he was born in Płońsk, near Warsaw (then under Czarist Russian rule), in 1886. It is, of course, an impertinence to deal with Ben-Gurion, who has been the captain of his people for more than thirty years, and is incontestably one of the great men of modern times, in a scant paragraph or two. He is a missionary-poet-statesman. In passing I might mention that a bibliography of his writing comes to three thousand items.

David Ben-Gurion, old as he is, still has punch and bite. He is as homely as a shawl, but with an austere intellect. His outstanding characteristics are, it seems to me, durability, courage, and acumen. His conversation is still peppered with lively aphorisms. He said to me in the course of a two-hour conversation, "The Jews chose God, not vice versa," and "The United Arab Republic is neither united, Arab, nor a republic." Maybe these are old cracks which he tossed off for the fun of it, but there can be no doubt as to the originality, sparkle, and comprehensive attentiveness of his mind.

Dedication, a gift for prophecy, single-minded hewing to the line—these are other qualities. Everything about Ben-Gurion is enhanced by his physical appearance. This is no stuffed or solemn father image; he looks like a cherub, with pink cheeks, an unlined face that seems somewhat compressed, a blunt nose, and stiff wings of pure-white hair. He has the brightest of bright blue eyes—children's eyes, which twinkle. Supposedly he stands on his head for half an hour every day, as did Jawaharlal Nehru. His handshake is soft, gentle. He wore a khaki army-issue shirt open at the neck, a khaki army belt, and tan slacks. His room was not air-conditioned, but he did not seem hot, although the temperature was high. He puts his hands over his eyes with a smack, then stretches them out, to indicate dismay when he tells a story.

A rough outline of his career goes as follows. He was one of eleven children; his mother died giving birth to the eleventh. His father was, strange as it may seem, "both an observant Jew and a free thinker"—these are his own words. Zionist meetings took place in his father's house, and young Ben-Gurion, or Green as he then was, heard passionate talk about Zion from the earliest age. "The Jewish religion is a nationalist religion," he explained. His father followed him to Palestine in 1925, with two brothers and sisters; all have died. His wife, Paula, an American nurse, whom he met on one of his trips to New York, and who was an exceedingly idiosyncratic and even anarchic character as well as his tenacious watchdog for many years, died last year. They had children, but Ben-Gurion lives alone. The loss of Paula was a major—almost mortal—blow.

Ben-Gurion decided to make his way to Palestine, under the Turks, against the wishes of his family, which was comparatively well-to-do. He arrived in 1906, aged twenty, and set out to work (like Eshkol) as an agricultural laborer, and also had a job for a time in a printing plant. He set out making Zionist propaganda at once, and the Turkish administration exiled him for a time, something that did not happen often. He taught himself French and English (decades later, he learned Greek in his sixties), and, returning to Palestine, identified himself strongly with the nascent labor movement. He served under General Allenby during World War I as a soldier in the Jewish Legion founded by one of the ablest and most controversial of the early Zionists, Ze'ev Jabotinsky, and became a member of the General Council of the Zionist

Organization in 1920. He was a major instrument in founding both the Mapai Party and Histradut, the labor organization, of which he was chairman for almost fifteen years, and he was the first editor and publisher of *Davar*, a labor newspaper which still exists. Presently (1935) he became chairman of the Jewish Agency, proclaimed the independence of the country on May 14, 1948, and served as Prime Minister and Minister of Defense from 1949 to 1953, then again after 1955. "To be Prime Minister was nothing much," he chuckled. "But to be Minister of Defense—that counted!"

Sometimes his gift of phrase and elevation above the battle is Churchillian. In 1939 after the outbreak of World War II when the British issued a famous White Paper halting Zionist emigration to Palestine, Ben-Gurion said, "We shall fight the White Paper as if there were no war on. We shall fight the war as though there were no White Paper."[5]

✼

I drove with Max Rosenfeld, a perspicacious young Israeli (born in Canada), who is both an official in the foreign office and chairman of Ein Karem, a village near Jerusalem well known for its associations with John the Baptist (peaches, almonds, grapes on Mr. Rosenfeld's terraced doorstep), to see Ben-Gurion at his kibbutz, Sde Boker, deep in the Negev Desert. Ben-Gurion also maintains an establishment in Tel Aviv, but he happened to be in the Negev at the time. Mr. Rosenfeld and I passed through Bethlehem, where the Church of the Nativity is a disappointing sight; through Hebron, where Abraham, Isaac, and Jacob are buried, a glass-making center now; and Beersheba (ceramics, tile, solar heaters), where Abraham dug his well. Part of this region was Arab territory before the Six-Day War. We passed Arabs driving in and out of the Gaza Strip. Arriving at Sde Boker (the name means "Field of the Ranger"), we set out to find the Ben-Gurion bungalow in a settlement in the middle of

[5] Catarivas, *op. cit.*, p. 99.

the desert. This desolate region lies north of what is called the Wilderness of Zin. The brilliance of the sun was penetrating. I don't think I have ever felt so punishing a heat except in Kano or Khartoum. I began to appreciate B.-G.'s durability. When he retired here, he worked outdoors in the dry burning slate of the desert as a shepherd, although eighty.

A paratrooper in a red beret with a Sten gun stood on guard outside the bungalow. The room where Ben-Gurion sat placidly awaiting us was solidly lined with books—classics and standard reference works in half a dozen languages. A big table was piled with skyscrapers of books as well. He chose this location partly because he loves it, partly because the desert and its possible cultivation are so important to the life of Israel. Here, with Kollek's help, he set up his University of the Desert, which consists of three bleak buildings, but which he talked about as if it were Oxford, Harvard, and the Sorbonne turned into one. Problems relating to the desert are studied in a research program. Ben-Gurion is the kind of old man to whom youth is attracted, and the young teachers and research workers were a lively lot. "When I came to Palestine, there were only twenty Jewish settlements in the entire country," Ben-Gurion told us. He did not need to enlarge on how this situation has changed. "Moreover, in those old days Jewish landlords used Arab labor—it was cheaper and more obedient." A transference in thought brought him to General Dayan, whom he loves like a son. "He, too, has this feeling about the desert. The whole West Bank is a desert, which cannot be irrigated or made to bloom because of the prevailing winds. Also, Dayan has a humane feeling toward the Arabs."

Principally Ben-Gurion elaborated on two interwoven themes, religious and political. Jerusalem must remain in Israeli hands united and undivided, because God promised the city to the Jews, but, if I understood him aright, he does not believe in annexation of the other occupied territories. Jewry has two aspects, he affirmed—"nationalist" and "universal." A Jew cannot be a member of any other nationality. "Suppose a German child comes to Beersheba to study, and becomes converted to

Judaism; after that he is no longer a German, but a Jew. He is a *universal* citizen as well." He returned to the theme of Jerusalem: "This is a Jewish country, and Jerusalem is our holy city," which is incontrovertibly proved by the Book. "Adam, the first man, was not necessarily a Jew, but 'Adam,' a Hebrew word, means 'Man.'" A human being cannot comprehend what is God. "God said, 'I will punish you, Israel, for your sins, because it was you, Israel, specifically, that I gave the Torah to,'" a sentence which I did not quite follow. Also: "The religion of the Jewish people is not the only religion of the Jews. The creator of Christianity was not Jesus, but Paul. Jesus was a splendid man, however—although we do not consider him to be a prophet, as the Moslems do. After Paul, Christianity was an easier religion to follow than Jewry. Islam is nearer to Judaism than to Christianity, because of its monotheism. Hence many Jews became Moslems in the fifth and sixth centuries A.D. Jews and Arabs are inextricably intertwined."

What worries Ben-Gurion most about the present political picture is that, he thinks, the Arabs have *not* learned a lesson from the Six-Day War. They are certain to attack again—try another round. "Why should they make peace? They outnumber us fifty to one." He added with grave emphasis that the Arabs can afford to lose battles, even wars, again and again, but they only have to win *once*, because an Arab victory would necessarily mean extermination of the Israeli state. "We have to win *every* time, but they do not." Solution? This, he believes, cannot be found except in the larger design of international affairs. He thought that a *détente* between the United States and Soviet Russia, which would relieve the latter of fear of China, would be an immense amelioration to the situation in the Middle East. "China is the reason why Russia cannot give a decent standard of living and cannot liberate."

The old man ended up by telling anecdotes, pretty good ones too, about Churchill, Weizmann ("He got us the Balfour Declaration"), Truman, Eisenhower, and Kennedy. His touch is always shrewd. For instance, he noted that Weizmann got along with

Conservative governments in England better than with Labour. He switched back to the present, and his last words were a gentle admonition to Max Rosenfeld, my companion—"You ought to change your name to a Jewish name!"—by which he meant that "Rosenfeld" should be Hebraized. First, last, and all the time, Ben-Gurion is an Israeli Jew.

13. A WALK ON
THE ARAB SIDE

The Jewish National Home was not, unfortunately, installed in a vacuum or empty lot, but in a region where Arabs had lived for more than fifteen hundred years. The Arabs, true, do not go back nearly so far as the Jews, and they had done comparatively little to build up the territory or make themselves a significant homogeneous force. Nor had they composed the government of the country, since it lay under the dominion of the Turks for centuries and then of the British. Most of the Arabs were, indeed, nomadic homeless Bedouin or peasants under effendi landlords. Nevertheless it cannot be contested that Palestine was "their" country, and this—only too obviously—is the root of the Arab problem in contemporary Israel and its resultant miseries.

The problem has been much exacerbated by the Six-Day War in June, 1967, when, as we know, East ("Arab" or Jordanian) Jerusalem was incorporated into the rest of the city, which brought in almost 70,000 Arabs. In Israel as a whole the Arab population rose by 750,000, as a result of the occupation of the West Bank, the Sinai Peninsula, et cetera, by the Israelis. This occupation may not be permanent, but it is germane to the present situation, if only because it promotes tension. The Arab population has, moreover, grown proportionately higher than the Jewish population in recent years, in spite of the large movement of Israeli Arabs into Jordan. Part of this rise can be attributed to natural causes; the Arab birth rate is higher than that of the Jews, and their death rate has gone down partly as a result of the admirable Israeli health services.

One of the principal Arab grievances is that, even though they comprise such a considerable share of the population, their children do not have a corresponding access to education; only a minuscule percent of the budget for higher education goes to Arab schools. There are only about 300 Arab students at the Hebrew University as against 7,500 Jews. The Israelis on their side say that this is not the result of discrimination, but comes about because few Arab students are equipped to qualify; the Arabs reply that they have had their early education in Arabic-speaking enclaves, and find it difficult to compete in an institution where instruction is in a different language, Hebrew.

Arabs are not permitted to serve in the army, if only because they might not be reliable, to put it mildly, in an Arab-Israeli war. Besides, the Israelis do not want to impose on them the possibility of their having to fight fellow Arabs. They do, however, serve in the police, and are screened for police service without discrimination. They have equal access to the courts, which are scrupulously fair, and they share in the general high standard of living in Israel, which exceeds that of any Arab state except possibly Kuwait and Lebanon.

Israel is usually thought of as a pepper pot full of contesting Arabs and Israelis in constant turmoil, but this is not necessarily the case. The Arabs are by and large a peaceable community. Some are still uneducated Bedouin, with little interest beyond their flocks and a bit of smuggling. There were scarcely any instances of Israeli Arabs making trouble of *any* kind, much less promoting insurrection or bloodshed, during the Six-Day War.

Terrorism on the frontiers is something else again, and has become a fiercely vexing problem, as every reader of a newspaper knows; but this seldom involves *Israeli* Arabs; bloodshed comes from thrust and counterthrust vis-à-vis Egyptians and Jordanians (less seldom, Syrians; still less seldom, Lebanese) along the country's all-Arab borders. A terrorist organization known as Al Fatah (the word *"fatah"* means "conquest"), a vigorously led and well-organized commando group, makes havoc where it can. Al Fatah also operates by way of a radio station, and broadcasts in Hebrew as well as Arabic. Its political orientation is leftist as well as being savagely anti-Israel, and it takes particular zest in attacking Ambassador Gunnar V. Jarring, the Swedish diplomat who is head of a UN mission in the area, on the ground that his basic aim is to confirm the present situation and "regularize" Israeli supremacy in the occupied areas.[1]

Many primitive Arabs moving into Jerusalem from the countryside have never seen running water, a traffic light, or even a park; digesting them (although "digesting" is not quite the correct word) presents difficulties. Most Israelis who dislike Arabs do not do so out of any consideration of race, language, religion (after all both Jews and Arabs claim a common descent from Abraham), or even politics, but because so many Arabs are culturally inferior, on a much lower social level. But here a counterpoint must be mentioned as well, namely, that a good many Jordanian and other Arabs in East *Jerusalem* (less so in the hinterland) have long records of education, cultural achievement, participation in worthy enterprises, and wealth. They do not take their new lot willingly, and some refuse to give cooperation of any kind to the Israeli authorities. The Israelis know this, and, as noted in a preceding chapter, do their best to make their administration inconspicuous, not out of solicitude for the Arabs but to keep the country running smoothly. There are extremely few Israeli administrators in East Jerusalem or in the 26,000 square miles of occupied territory, probably fewer than two hundred in the whole country.

Israeli Arabs cover a wide range, from distinguished lawyers and professors in the cities to ghostlike penurious refugees in the hinterland. The Mayor of Hebron, an Arab Sheik named Ali Mahmoud Ja'abari, is one of the best public officials in the country; so is the Mayor of Nablus. Predominantly Arab towns like Jaffa, Acre, Nazareth, Afula, Ramla, Lydda (the seat of the

[1] Dana Adams Schmidt in the *New York Times*, September 27, 1968.

airport) are essential constituents of the Israeli state; their inhabitants cannot be ignored. In general, the Jewish and Israeli
Arabs get along together fairly well, and often work side by side
in offices, shops, and so on. There is no tension at all to compare
with what existed during the mandate, when the country ex-
ploded into savage riots (1929), particularly in Hebron and Safed.
In those days the Arab leader was a brazen marplot named Haj
Amin el-Husseini, the Grand Mufti of Jerusalem. He worked with
the Nazis during the war. I remember those riots almost thirty
years ago quite well, because I was there.

It is the Arabs who, I heard one bright young woman say, are
the real Zionists today. It is they, rather than the Jews, who seek
to return to their homeland and hope for permanent and secure
settlement. And one must never forget that Jerusalem, the city, is
just as holy to Moslem Arabs as it is to the Jews and Christians.
It was from the Haram esh-Sherif, after all, that Mohammed, firm
on his winged horse, rose to heaven.

One youthful Moslem journalist whom we met emphasized
that there are *two* "just" causes in this country, a statement that
not many intolerant Israelis would agree with. When I asked
about solutions, a number of ideas came up, none of them very
practicable at the moment. One is that Jerusalem might be given
a new administrative status and become a double capital, first of
Israel as a whole, second of an autonomous "Palestine" in the
predominantly Moslem areas. There might even be a "third"
Jerusalem which would be the capital of a Middle Eastern Customs Union.

The principal Arab grievances are these—all deriving from the
major overriding concept that the Israelis had no right to have
established their National Home here at all, since this is an Arab
country:

First, the fact that the Jews have acquired large tracts of land,
and their refusal on occasion to employ Arab labor. (But the
Israelis say that the Arabs sold the land to them in the early days,
often at a handsome price.) Second, Jewish immigration. Third,
large numbers of Arabs are still held in the refugee camps in the

Gaza Strip and elsewhere. (But the Israelis state in rebuttal that this came about as the inevitable result of warfare, that most Arabs in the camps are unemployable, that they are not prisoners but are free to leave the camps at will, and that they would constitute a serious security as well as economic problem if the camps were abolished.) Fourth, funds are inadequate for Arab education. (Here, the Arabs certainly have a point.) Fifth, they are treated as second-class citizens in general. Sixth, the Arabs in annexed East Jerusalem have an anomalous status. "They are citizens of Israel, yet nationals of Jordan. They possess full municipal rights and obligations, including the right to vote and put up candidates in the city elections next year. Yet they are denied— nor have they claimed—citizenship in the state of Israel in which they now reside."[2] Finally, the Arabs fear outright annexation of the occupied areas.

The Israeli answers to these charges are complicated. A good deal of hysteria gets into the issue (on both sides). The Israelis deny that they are interlopers or later arrivals; a firmly rooted community of Jews has lived near Nablus, as an example, since 722 B.C. The Arabs held the country only from A.D. 637, when the Ommiad califs, who were not Palestinians at all, began to rule in Damascus, until the country was invaded and seized by the Ottoman Turks in 1517. Moreover, Palestine was never an independent Arab country, although it is incontestable that, even under the Turks, the population was overwhelmingly Arab. There was, however, no nationalist movement to speak of until just before World War I, when this was stimulated by the British. Most land was in the public domain, and the early Zionists had every right to buy it. Arabs today own substantial amounts of land, and many shops along such a thoroughfare as Jaffa Road in New Jerusalem have Arab landlords. As to education, Arabs have their own Arab teachers in their own language in their own schools, exactly as if they were still living in Jordan. Jewish standards of education are, however, higher. The Arabs are *not* second-class

[2] *Economist*, London, July 12, 1968.

citizens, the Zionists insist, by any reasonable definition of the term. The individual Arab may not feel at ease, since this is a Jewish country, but he is not harassed. The proof of the pudding is in the eating. Only about *twenty* Arab families or businesses have left Old Jerusalem since it became incorporated into the new. As to this last point, the Arabs say in rebuttal that it is a matter of honor for their leaders to remain, and they stick it out as a symbolic gesture.

Intermarriage between Jew and Arab is uncommon; there are more cases of Arab boys marrying Jewish girls than vice versa. Arab girls are kept in purdah if their upbringing is strict, and their marriages are arranged as a rule by the parents. But Arab boys have more opportunity to mix with Jewish girls than before. Oddly enough, two of the most conspicuous Arab terrorists caught recently have Jewish mothers. A considerable diminution of hostility between the Arabs and Jews has come in Jerusalem since the June war. The two communities meet and mingle on a scale hitherto unprecedented. More and more do business together. Jerusalem is a kind of "pilot" plant for the future in this respect. Arabs are beginning to learn Hebrew, and progressive young Jews take group lessons in Arabic. The Jews have lost their terror of "savage" Arabs marauding into the city, and the Arabs have discovered that most Jews do not have horns or tongues of flame. The heart of the problem from the Jewish point of view is not that the Arabs are warlike and dangerous, but that, as indicated above, they seem to be living in medieval times.

We talked to several Moslems, Palestinian and other. Abdel Aziz Zwabi (more correctly, "Z'ubi"), one of the Arab members of the Knesset, has been both Deputy Mayor and Mayor of Nazareth; he is one of the editors of *New Outlook,* an interesting monthly published in English in Tel Aviv, which takes a strong conciliatory line in Israeli-Arab affairs. Mr. Zwabi is an outstanding member of Mapam, the left-wing opposition party which is the only Jewish-Arab party left in the country. There can be a bit of confusion here; another member of the Knesset has the identical name, and Zwabi's brother, a more conservative political figure, has also been Mayor of Nazareth. And a brother-in-law is a cabinet minister in Jordan! So can members of the same family be mixed up in this part of the world. "Some Arabs vote for Jews, but no Jew ever votes for an Arab," Mr. Zwabi told us, "but as for myself I will agree with a Jew, rather than an Arab, if I think the Jew is right." In his opinion most Arabs have now come at long last to recognize that Israel is a fact, a reality, and is here to stay. The problem has, he thinks, become bigger in recent years; there is no longer a "Palestine problem," but a "Middle East problem," with areas a thousand miles away acutely involved. "The Arabs are a people of emotion; that is why we always lose. We think with our hearts, not our minds, so that we lack skill, experience, unity." He mentioned that Saudi Arabia and Jordan had not been on good terms for forty years, that Lebanon fears Syria, and Iraq doesn't get along with Egypt. There are, he thinks, several keys to a peaceful settlement between Israel and its neighbors, the chief of which is in the hands of the Israelis; this would be a formal declaration that it has no intention of further expansion. Many Arabs think that the fundamental aim of Israeli policy is territorial aggrandizement all the way to the Euphrates.

Mr. Zwabi is a robust person, good-humored, with a direct, warm, open smile. He has served in the Knesset for three years. He took us for a tour of the building, and he must have been stopped a dozen times by fellow deputies—Jewish and other friends—who seemed to be gathering to themselves something of his own optimistic vitality.

One day we called on Anwar Nusseibeh, a distinguished Arab lawyer and public figure, Cambridge-educated, at his home in the Old City; he has been variously Jordanian Minister of Defense, Minister of Education, Ambassador to London, and Governor of Jerusalem under Jordan. Unlike Mr. Zwabi, he not only refuses to take part in public life under the Israelis, but boycotts it. He has even given up his law practice. His loyalty to Jerusalem the city, *his* Jerusalem (the Old City), is, however, undiminished. He thinks, in fact, that the UN might well move from New

York and establish itself here instead, since Jerusalem is "the center of the world." Mr. Nusseibeh was both calm and logical in his carefully discriminating statement of Arab grievances; mainly these had to do with what he called injustices in the field of education. A minor annoyance is that, nowadays, he and his compatriots have to go through a good deal of red tape and ask permission of the Israeli authorities in order to go to Amman (the capital of Jordan). Formerly, of course, this trip was simply a matter of getting into a car for an hour's ride, since East Jerusalem lay *in* Jordan. "We have no quarrel with what the Jews want to do with *their* Jerusalem," he told us. "All we ask is to be left alone with ours. After all, Jordan retained Amman as its capital when it was created as a state; it didn't try to come here." Later Mrs. Nusseibeh joined us; she was passionate about what she called the other "injustices" wrought by Israel. Seldom have I been conscious of so much flame in a voice so cultivated.

Another interesting and talented person is Abdullah Schleifer, a tall, youthful American of Jewish origin who became a convert to Islam some years ago and who lives in Old Jerusalem with his pretty American wife, a Negro, who is also a Moslem by conversion. They make an exceptionally attractive couple, resolute in their convictions but capable of moderation and an objective view. Schleifer, a journalist and poet, edited an English daily newspaper on the Jordan side until the 1967 war closed him out. At present the city has no newspaper in Arabic except *El Yom,* a propaganda journal published by the Histradut, the Zionist labor organization; before the war there were three, but none in English except Schleifer's.

Mr. Schleifer gave us his views on some background matters. "The Jews," he said, "use religious mythology as a basis for secular nationalism"—a succinct way to express this point. In reality, Schleifer went on, such a revered Jewish hero as David was probably no more than an anonymous Bedouin "out of some Semitic tribe." The Arabs of today, he proceeded, have no wish to destroy the Jews as such; what they do want to destroy is the Jewish state. The Palestinian Arabs are, he said, by far the most educated and competent of any Moslems in the Middle East, except possibly those from Lebanon; they are the administrators, the teachers, the solid basis of the civil service, all the way from North Africa to Iraq. What the Israelis are trying to do in Jerusalem above all is prove that Jews and Arabs *can* live together. He conceded freely that Israel was a democracy for the Jews, and said that the Israeli community has become, so to speak, solidified; conspicuous anti-Zionist Jews have become absorbed by Zionism. He listed several Arab grievances. One was that the Jerusalem Moslems had no political party, but then he proceeded in all fairness to add that Jordan did not have a political party either. Five or six Arab leaders (in the Jerusalem sector) have, he said, been exiled by the Israeli Government—banished to other communities—and several have been forced to move to Amman. Several sizable towns, thought by the Israelis to be incorrigible nests of Arab terrorism, have been wantonly destroyed—burned to the ground. Finally, there are, in his view, crying injustices in the way Arab property, like shops on the Jaffa Road which were seized and taken over by the Israeli Government after the Six-Day War, is being administered under circumstances close to confiscation.

✶

To sum up, the central dilemma is insoluble. The Israeli majority cannot possibly assimilate the Arab minority, even if this were socially or politically feasible, because in that case Israel would cease to be the Jewish National Home. If integration takes place, the country is no longer a Jewish state. If the Jews dilute their concept, they will lose it. So the Arabs have to be dealt with as a permanent unassimilable minority, since they, on their side, are not strong enough under present conditions to take the country over.

Meanwhile explosions take place daily along the frontiers—terrorist raids, bombing, exchanges of gunfire, with the result that the whole of the Middle East is aquiver. The great powers quiver too, with the United States on the Israeli side, more or less, the Soviet Union on the other. The Soviets are arming the Arabs hand over fist, particularly Egypt; the United States sells military jets to Israel. It is painful to have to conceive that a conflict brought about by disruptions in Jerusalem, the City of Peace, may quite possibly bring on the Third World War.

14. "PEARL OF THE
MIDDLE EAST"

Beirut commits treason against itself. This ancient city, the capital of Lebanon, blessed with a sublime physical location and endowed with a beauty of surroundings unmatched in the world, is a dog-eared shambles—dirtier, just plain dirtier, than any other city of consequence I have ever seen. While I was there its 600,000 people had their water supply cut off for ten hours one day because a man committed suicide by plunging into an open drain in the main conduit of the city's water supply.

In the best quarter of the town, directly adjacent to a brand-new hotel gleaming with lacy marble,[1] there exists a network of grisly small alleys which, so far as I could tell, are never swept at all. Day after day I would see—and learned to know—the same debris: bent chunks of corrugated iron, broken boulders or cement, rags, rotten vegetables, and paper cartons bursting with decayed merchandise. Much of the detritus seems to be of a kind that goes with a rich community, not a poor one. The cool and sparkling Mediterranean across the boulevard looks enticing for a swim, until you see that the water is full of orange peel, oil slick, blobs of toilet paper, and assorted slimy objects, which it is better not to look at or try to identify.

I have a friend, a diplomat, who lives in a neighborhood called Ain Mreisse, the equivalent of the East Sixties in New York, and who has experienced a good deal of what he calls somberly "the

[1] And designed by no less an architect than Edward Durell Stone.

sewer complaint." A nearby main sewer discharges its flow onto the Saint-Georges Hotel beach, polluting it, or causing havoc upstream, depending on the currents or wind. A building housing a handicraft display, owned by the government, was built directly above this sewer three years ago and has never been able to be used, on account of the stench. My friend has a keen interest in another smaller sewer as well. It erupted into the garden of the house next door, ran down to the sea, and polluted the nearby area. My friend and his neighbors protested to the municipal authorities. No result. They signed a round-robin petition and presented it to the Minister of Health. No result. So eventually they bought some cement and plugged the sewer themselves. But this, unfortunately, caused it to back up and flood the main street of the district. The government, impressed by such an unheard-of example of initiative, finally took action—excusing its previous inattention on the ground that this was a *private* sewer belonging either to the embassy of a great power nearby or to a Kuwaiti businessman. The action the government took was to *remove* the cement plug my friend and his fellows had installed. They put it in again, it has stayed in, and the malodorous problem is solved—at least for the time being.

Lebanon has ski runs in the mountains north of Beirut. A traveler reported to me, "Even the snow was filthy—a dirty yellow. Picnickers left garbage on the runs."

Landlords in Beirut are, to put it modestly, not all that they should be. A member of the Marine guard at the United States Embassy now walks up the four flights of stairs to his apartment. Decision to do so came after he had been caught in the elevator *between floors* seven times.

The currency is by and large so filthy that it feels fungoid and moist—like the paper money in primitive Balkan kingdoms before World War II. It is almost a necessity to wash your hands after handling any money—if water is to be found.

I would be interested to know what the standards of medical practice are of a doctor whose name plate in a centrally located

office building says, "SPECALSATION IN HARD MEDICAL CASES." Next
to him is the "007 Spy Institute."

There are streets in the slums blocked with filth, including
donkey dung, that would not be tolerated for a second in Senegal,
Cawnpore, or even the old Jersey City under Mayor Hague. Or
think by contrast of Jerusalem, where a person could eat dinner
off almost any street. But—let us remember!—at night the Beirut
shoreline is strung with pearls shining in the hills, winking along
the shore, under a teardrop moon. Illuminated crosses, as in
South America, shine on mountaintops. The backdrop to Beirut
is as incomparably lovely and thrilling as the city itself is slovenly.
Recently—an odd irony—came the threat of a strike by—of all
people—the street *cleaners*. Last year a similar strike of four
thousand sweepers "jeopardized health conditions" for several
months.

Another manifestation of civic decay has to do with the public
services—telephones, transportation, water supply, and so on. As
of a recent date there were only two public telephone booths in
the entire city.[2] About 120,000 children of school age do not
go to school because there are no schools to go to. Still one
more indication of decay is, as goes almost without saying, corruption. Nothing is so minor in Beirut that it may not be cor-
rupted. This is the capital not merely of Lebanon but of that
mysterious entity known as the Levant, and here are to be found
expressions in several fields of the concept known as "Levantine"
in the most opprobrious meaning of the term.

Rule by Ratio

The curse of Lebanon, an extraordinary complex small country, is sectarianism, and the people are governed by what is
known as the "Confessional Balance," which means that the basic
stratification is not by political party, but by religion. There are
no fewer than fourteen different religious sects in the country,

[2] Thomas F. Brady in the New York *Times*, November 20, 1965.

rival members of which fight like ocelots from time to time. On an over-all basis the population is roughly 50 percent Christian, 50 percent Moslem. A considerable number of Jews live here as well, perhaps six or seven thousand, but they do not count much in the realm of politics. Jews are, by and large, neither discriminated against nor harassed. When the Six-Day War broke out in Israel in June, 1967, the suggestion was made by Christian groups in Lebanon to evacuate the Jews from Beirut for their own safety, but this turned out to be unnecessary. There were no anti-Jewish disturbances of any kind.[3]

The Confessional Balance is a political curiosity of considerable interest. It rose out of what is called the "National Pact," which was devised some years ago to ensure political stability in the country and which has underpinned it ever since. In essence, the pact was a compromise or tacit bargain between the two main divisions of the nation, whereby the urban Moslems gave up their long-standing desire to unite politically with the Moslem hinterland, and the Christians surrendered their hope of a continuing occupation by the Western powers, which would protect them. So a ratio close to 50-50 was set up. As a result—one picturesque result among many—the rule was established that the national parliament must contain six Christians (of various denominations) for every five Moslems. This means in turn that its membership must necessarily be a multiple of eleven—once there were sixty-six members, now ninety-nine.

Furthermore political offices are divided rigidly among the various religious communities in a manner unknown elsewhere in the world. The President of the Republic must always be a Maronite Christian, the Prime Minister a Sunni Moslem, and the Speaker of the House a Shia Moslem. (In case this seems startling, one may recall that an "ideal" electoral ticket in New York consisted in older days of one person of Irish descent, one of Italian descent, and a Jew.) The commander in chief of the army in Lebanon must be a Christian—preferably Maronite—and the chief

[3] The Lebanese Government sat out the war with gingerly care, and did not join in the fighting on the side of its Arab neighbors.

of police something else. Army and police do not, incidentally, get along too well together. The Foreign Minister must be a non-Maronite Christian, and the Mayor of Beirut a Greek Orthodox. The Christian membership of parliament consists of thirty Maronites, eleven Greek Orthodox, six Greek Catholics, four Armenian Orthodox, one Armenian Catholic, one Protestant, and a few others. The Moslems are divided between twenty Sunnis and nineteen Shias. Of course, in order to rule smoothly members of the various communities make incessant deals with one another. Patronage and jobs are given out strictly on a *religious* basis. The first question asked of anybody is always, "What is your religion?"

Driving to Baalbek one day, I asked our driver, "What are you?" and he replied sharply, "Lebanese." He proceeded to break this down by adding first that he was a "mountaineer," second that he was a Greek Catholic. The fact that the driver first used the word "Lebanese" meant, to those who know, that he must be a Christian. A Moslem would have replied that he was an Arab. On the other hand, some Arab citizens might expand this by saying, "I am a Moslem first, a Lebanese second, and an Arab third." Our driver's reply had, I was then told by my companion, both a positive and a negative connotation. Positively, by saying "Lebanese," he meant that he was expressing a sense of the country's nationalism. Negatively, he was making it clear that he was not an Arab. There are plenty of *Arab* Christians in the country, i.e., men and women of Arab stock whose ancestors became converted to Christianity long ago.

The cabinet, as of April 10, 1968, contained three Sunni Moslems, three Maronites, one Shia Moslem, one Greek Orthodox, one Greek Catholic, and one Druse. "Everybody is a minority; there is no majority, and this is the key to everything," one neutral diplomat told me. A document prepared by a Western embassy paraphrases Churchill: "Lebanon is a compromise wrapped in a tacit understanding and mounted on gimbals."

As to the Moslems, it is scarcely within the province of this book to deal with the differences between Sunnis and Shias, a subtle and complicated subject having to do mostly with the hierarchical position of Ali, the fourth Caliph after Mohammed; suffice it to say that the Sunnis are the more orthodox. But about the Christian Maronites we must have a word, because they have been so conspicuous in the history of the region from early times, and play such a dominant role today. As a matter of fact, precise definition of the term "Maronite" is not altogether easy. Maronites are, to put it as simply as possible, Roman Catholics who accept the supremacy of the Pope, but whose litany is in Arabic, which brings them close to the Greek Catholics. They are the only Eastern Christian church which never broke with Rome, and one conspicuous Cardinal in the Roman Catholic Church today is a Maronite, His Beatitude Cardinal Paul Meouchi, resident in Lebanon.[4] The Maronites sometimes use the Assyrian language as well as Arabic in their flowery ritual. The backbone of *French* secular power in Lebanon in the modern period has always been the Maronite Church. The Maronites were inextricably involved in French political maneuver during the period when the French held the mandate for Lebanon and later, and many upper-class Maronites speak better French than Arabic; most are bilingual. The *British* as a rule tended to consider the Maronites to be "Non-Arab," but this was a contradiction in terms. Balancing French support for the Maronites, which was their rock-bottom policy, was Russian support for the Greek Orthodox. The French resorted to every contrivance for year after year to keep Maronite influence supreme. They said, in effect, "You will be favored like us if you are as French as we are." Anything Gallic was acceptable.

The Maronites are called "Uniate" because they have never, in the words of a private memorandum, "been properly separated from Rome." The Greek Catholics are former Greek Orthodox "who split off from the mother church in 1709," and the two communities still have several important points of difference.

<hr />

[4] He is also known as "Patriarch of Antioch and All the East." The present Patriarch was, of all things, at one time an American citizen, and spent fourteen years in the United States.

Similarly, "the Syrian Catholics (1667) are an offshoot of the Syrian Orthodox Church, and the Armenian Catholics (1740) of the Armenian Orthodox." Syrians and Armenians, like the Maronites, are permitted to retain the use of Syriac ("an Aramaic dialect which was superseded by Arabic in the 13th century") and Armenian. There are also Assyrians, known as Chaldean Catholics, who follow a variety of dogma known as Nestorian.

To sum up a complicated matter, the "Uniate" churches are the Maronite, Greek Catholic, Syrian Catholic, Armenian Catholic, and Latin (Roman Catholic), numbering about 536,000 communicants. The "separated" churches—Greek Orthodox, Syrian Orthodox, Armenian Orthodox, and Nestorians—number about 262,000 members. There are 13,000 Protestants. Some Greek Orthodox switch over to Greek Catholic on occasion.

The Arabs, no matter of what subdivision, and whether they are Christian or not, often give a down-at-the-heel impression. Arabs who are active politically and who take sides emotionally with Colonel Nasser in Egypt, as most do, are known as "hot" Arabs. One contentious remark I heard was, "An Arab in Lebanon is a Jew who needs a shave."

The most outstanding community in the country is probably that of the Druses, whom we have met in preceding chapters, and who are an offshoot of a Shia sect known as Ismailis. Druses are by and large proud, clean, and honest. No Druse women ever become prostitutes, and if a young Druse girl goes astray, her father or brothers will kill her.

Politics are fiercely interwoven, and, to revert to a theme already stated, there are multiform opportunities for corruption. A recent issue of an English-language Beirut daily mentions that Dr. Bakhos Hakim, the deputy for Koura, put a question to the government before parliament in regard to the asphalting of certain private roads and courtyards belonging to two men "whose brother is a certain judge we all know." He names the names. Other violations, according to Dr. Hakim, took place in the village of Btoritage, "particularly the asphalting of private roads and courtyards belonging to the father and brother of a certain security man." On June 15, 1968, another paper describes how Deputy Ahmad Isber, a member of the House Office, "was surprised to discover a stranger having an unauthorized fifteen-minute telephone call with Senegal." On inquiry it was discovered that "the man was not a civil servant and his conversation with Senegal had nothing to do with parliamentary affairs." Another report by Deputy Abdo Saab said that "more than fifty civil servants in the House were getting substantial salaries 'but doing nothing.'"

Word about the Country and the City

Even geographically, Beirut is divided into distinct religious areas, closely set together so that mosques rub stones with Christian churches. There are specific Armenian, Jewish, Greek Orthodox, Maronite (three), Shia Moslem, and Sunni Moslem (four) quarters. The city observes two holidays a week, Friday for the Moslems, Sunday for the Christians. Nobody knows what the exact population is, because no census has been taken since 1932; the reason for this is that the Christian communities, knowing that the Moslems breed faster, want to keep the figures under wraps. The conventional guess is around 600,000, about a third of the population of the country as a whole.

Some Beirut residents are fascinating types. There are all manner of picturesque people, from merchants in the gold bazaar, whose financial minds go quicker than an abacus, and who have an unparalleled capacity for skillful and dramatic bargaining, all the way to taut slender young men in the fashionable bars fingering their amber beads and stout pashas in red tarbooshes smoking water pipes in the boulevard cafés. A good many women are still veiled, even though this is a liberated, advanced, and pro-Western country. One day I saw a bearded priest in a long black robe and black tarboosh, with a cigarette dangling from his full lips protruding above the beard. Fat Arabs with sagging bellies spit, spit, spit along the crowded sidewalks.

Generally the mood of the city is relaxed, in spite of the vulpine greed of the money-makers and the fierce noise of traffic. Anybody unusual in appearance, or too usual, is thought to be a spy. A friend in the Saint-Georges bar, a basic gathering place for the "in," once pointed out a well-known figure to me, describing him as "a spy for all seasons."

History goes far back, as it always seems to do in this part of the world. "Lebanon" means "white," and the region—"the Switzerland of Asia"—won this name because of the glorious peaks striped with snow, like couchant tigers, rising from the shore. Bold Phoenician sailors and traders began to use ports in this small country, like Tyre and Sidon, at least three thousand years ago; among other things they founded Carthage and penetrated over unknown seas all the way to Cornwall, where they mined tin. Solomon's Temple is supposed to have been built out of the great cedars which, rising to the north of Beirut, are still one of the sights of the country, and St. George, the patron saint of the nation, is supposed to have slain his dragon on the site of the Al-Khodr Mosque. This was the fabled "land of milk and honey," and some historians say that the first alphabet was invented here. But the principal distinction of Beirut is that, now as well as in olden times, it was a meeting place, a forum, for both East and West, equally open to the Mediterranean world, Europe, Asia, and the East.

Various conquerors came in the familiar pattern; it took Alexander the Great longer to take Tyre than to conquer all of India, because he had no fleet and the stout Tyrians controlled the sea. Then arrived the Romans; two professors of law in Beirut were among the authors of Justinian's Code. The Arabs or Saracens, the Crusaders, and the Turks followed in time; but, the Lebanese say, their country managed to preserve its own special and important characteristics right through to today. A paramount factor in this was the towering parapet of mountains above the beaches, in consequence of which groups of refugees, including the early Maronites, were able to hide, make pockets, and in some areas build secret monasteries.

Ottoman rule lasted from 1517 to the end of the World War I. But the wily Lebanese were never totally extinguished by the Turks, and French influence began to be decisive as the Ottomans decayed. Napoleon III sent an army here to protect the Maronites, and in 1864 a protocol was signed guaranteeing Lebanese autonomy under the protection of the European powers. The Turks were obliged to appoint a Christian governor in Beirut, under foreign control, and the Lebanese were exempted from paying taxes to Constantinople and from serving as conscripts in the Turkish Army. A pre-World War I guidebook makes note of the fact that Egyptians, Austrians, French, Russians, and Italians all had their special stations in the customs-house.

The contemporary period is marked by wild confusions and aberrations, which it is impossible to go into in this space. The French and British made a secret deal, the Sykes-Picot agreement, in 1916, for the future administration of the region, dividing it into spheres of influence. The British took over Palestine and Trans-Jordan as a result, and the French created a monstrosity known as the Grand Liban in what is now Syria and Lebanon. As always, they sought to rule by fragmentation. French policy in the Levant (as in Indochina) was to decentralize in order to break up local nationalism and make their puppets easier to handle. The Grand Liban was subdivided into five different substates, including Syria and Lebanon, each with its own constitution, flag, et cetera. Had it not been for this, Syria would probably be part of Lebanon today, or vice versa. The French cut Lebanon down to a sliver. Their rule, which was arbitrary in the extreme, lasted roughly from 1920 to 1943. I went several times to Beirut and Damascus in those days, and I have never seen such a concentrated expression of colonial rule before or since. The French treated the people like rats, even though they were supposed to be operating under a League of Nations mandate. The Druses made a heroic but unavailing revolt in 1926. Finally, during the early days of World War II the French were forced out after a weird little war with the British. It was in this war that General Moshe Dayan was wounded, as was mentioned in a preceding chapter. "The British were very polite, they never bombed us except at 9:15 A.M. and 4:30 P.M., just after tea," a Lebanese told me. The country was administered for a time by a British proconsul, the well-known General Sir Edward Louis Spears, whose wife, a novelist of distinction under the name Mary Borden, was Chicago-born and is the aunt of the former Mrs. Adlai Stevenson. Finally, the British left the country, and Lebanon became an independent state in 1943. A minor episode bringing it to the attention of the world occurred in July, 1958, when President Eisenhower ordered U. S. Marines into the country (on the invitation of the Lebanese Government) to ensure tranquillity after an armed insurrection, probably fomented by Egypt, broke out against the pro-West government of Camille Chamoun. Mr. Chamoun is still a well-known political figure. An attempt to assassinate him came in 1968.

✿

A curiosity is that more Lebanese live outside the country than in; this is true of no other nation in the world. Large numbers of Beirutis and other Lebanese have emigrated and set up colonies all over the world for many years. There are about 113,000 Lebanese in the United States, 150,000 in Argentina, more than 300,-000 in Brazil. One should mention, too, that the country has a substantial tourist traffic, and that Beirut is visited by more than 600,000 people from abroad in an average year.

✿

I have said that the inner mood of Beirut is relaxed, but it has a great deal of clatter and bustle. It has grown fastastically in the past few years, with a consequent building boom. What is now a moderately handsome residential section a mile or so down the Corniche (Rue 1) from the Saint-Georges was nothing more than a series of empty lots, a wilderness, ten years ago. Real estate prices are as high as—or even higher than—in New York. A small apartment in a good new building can cost $1,000 a month, and services are expensive as well.

One reason for the building spree is the influx of rich Kuwaitis. It is untrue to say, as is sometimes said, that Beirut has become a kind of exclusive skyscraper tent for Kuwait oil millionaires, but a good many rich citizens of Kuwait and similar small Arab states do live here, or come to the highlands for their holidays. The Kuwaitis chose Beirut over Cairo, as an example, partly because Egypt is too explosive politically, partly because Cairo lies in the sterling area. Most Kuwaitis do not trust sterling, and many lost large sums when the British devalued the pound. Beirut was safer. They could get their money out. Moreover, the Beirut banks were much more suited to the Arab taste than the old British-influenced banks in Cairo. In Beirut a Kuwaiti visiting a local bank is welcomed in Arabic fashion and made to feel comfortable—even given cups of coffee or mint tea while doing business. The Kuwaitis are a proud and touchy people. Mostly they put their money in real estate, and they prefer to leave a new building empty rather than lower the rent.

Another social phenomenon is the gradual erosion of French influence and the rise of American influence. The French gerrymandered the community with the aim of confirming and maintaining Maronite predominance, but this effort has not been altogether successful. Plenty of citizens today, especially the youth, do not speak French at all, whereas its use was formerly universal; English is gradually replacing it except in the solidly Maronite districts. Many Beirutis are bilingual today in Arabic and *English*, not Arabic and French. The two "cultures" center around the rival universities, and this, too, gives the use of English an edge, because of the immense prestige carried by the American University of Beirut, or AUB.

There are five universities in Beirut, which is one reason why Lebanon has the best literacy rate of any Middle Eastern country except Israel. Of these the foremost is unquestionably the AUB. Founded in 1867 by American Protestant missionaries, this institution has become famous everywhere in the area—indeed, all over the world—for its high academic and other standards and its devotion to the best in undergraduate instruction and subsequent training in the professions, like medicine. It had originally a faculty of 7; today, 700. About 3,500 students, both boys and girls, representing 59 different countries and 25 religious groups, attend its classes on a splendid campus. The AUB is the most stimulating and rewarding sight in Beirut. The French University, known as St. Joseph's, is likewise about a hundred years old, and was founded by the Jesuits. The other three universities are the Lebanese National University, set up under UNESCO auspices, the Beirut College for Women (672 students from 40 nations), founded by American Presbyterians in 1924, and the Arab University, financed in part by the United Arab Republic. It has approximately 11,000 students.

Beirut has fifteen beaches, of which several are spectacularly beautiful as well as agreeable, and forty-nine daily newspapers in Arabic, an astonishing number, as well as four in Armenian, three in French, and one in English. No country club for the rich exists, probably on account of sectarianism; the elite have to entertain at home, and so men are seldom free of their families. There are no sights in the sense that Jerusalem has sights, but it might be noted that the parliament stands in the center of what was the old Roman city. Automobiles go fast—crazily fast—and the briefest taxi ride can be an adventure, but accidents are comparatively rare—only forty-six fatalities in 1967. Drivers carry objects colored blue to ward off the evil eye. Streets are numbered for the most part Rue 55, Rue 81, and so on. The most fantastic private house I have ever seen is that of Madame Linda Sursock, the last *grande dame* of the local society; it contains pools, fountains, Ispahan rugs to be donated to the British Museum, and autographed snapshots of Noel Coward.

Communications in Beirut are by and large terrible; to try to telephone Paris is like trying to reach the moon. The Kuwaitis are forced to communicate with their homes by a device like ham radio. Creative writing and expression in music and art are minimal, but several political cartoonists are clever. Mail, even that of casual visitors, is likely to be opened by the police, although this will be denied.[5] The airport is an oddity. Beirut lies almost exactly halfway between London and Tokyo, and is an essential fueling stop for intercontinental jet traffic. The glide path for South Runway No. 18 lies directly over the public golf course, and the wits say that incoming planes make an undue hazard for golfers.

His Excellency the Administrator

I called one morning on Chafik Abou-Haidar, the principal executive officer of the local government, who is in effect both mayor of the city and governor of Beirut Province, one of the five provinces in the country. His name sounds Moslem, but he is Greek Orthodox. Mr. Abou-Haidar is a stocky, rubicund man with spectacles, heavy jowls and a pleasant, somewhat florid manner. He wore a large dark red ring on a thick finger. With us sat Georges J. Riachi, the City Engineer, clearly a well-educated and able citizen. The anterooms in the big, yellow, semiornate palace of the municipality on Yousef Rami Street were, in the Arab manner, full of lolling supplicants and retainers who sat inertly on benches for hour after hour. I did not notice at first that a young man in military uniform, but apparently unarmed, stood in the doorway just outside our view while we talked—silent, attentive, obviously a bodyguard.

The Administrator of Beirut is appointed to his post, not elected. He serves an unlimited term after being chosen for the job by the Minister of the Interior. Mr. Abou-Haidar is a lawyer by profession, aged fifty-six. He became a judge early in his career, and served as such for twenty-three years; he is not a professional politician. His official title is "Mohafez," if I have transliterated this accurately from the Arabic, and his speech is typically Arab—throatily soft and guttural.

Mr. Abou-Haidar's conversation covered a wide range of topics

[5] A story complaining about censorship of the mail appeared in the English-language newspaper on June 15, 1968.

for an hour. We met early in the morning, since office hours in Beirut run from 8 A.M. till 12 noon, no longer. His principal preoccupations—as with practically all the mayors we have encountered in this book—are population pressure, housing, and traffic. I asked him why the maintenance of public services is so extremely inadequate in Beirut—and he replied that these matters were not within his province, but belonged to the national government. The population of the city is bigger by day than night, he told us, since people swarm in from the suburbs to work, and this naturally links into the traffic problem. There were 1,072 "residential units" in Beirut in 1945, more than 56,000 now. This increase is so large that facilities for sewage, parks, utilities, and so on are difficult to maintain. The area for licensed building has increased fourteen-fold in fifteen years. A plan is projected for the city's future, but the problem is made difficult by Beirut's geography; the sea cuts it off on one side, the mountains on the other, so that no expansion is possible except to the north and south.

Juvenile delinquency is not yet widespread, nor is actual crime. But this is a *European* city, the Administrator stressed, with multitudinous temptations. Hashish is grown everywhere in the countryside. A lot of girls are beginning to wear miniskirts. As to traffic, this, he chose to say, was "informal," by which he meant chaotic. Automobiles choke narrow winding lanes on steep hills, and there are starfish intersections as in Rome. The Administrator asked himself a hopeless rhetorical question: "How can one control traffic here without rebuilding the entire city?" Finally, he talked about the special position of the minorities, in particular the Jews. "The Jewish community here is strictly a religious unit, which has no relation to Israel at all."

When I saw Mr. Abou-Haidar, he was still in his first year of office as Administrator. Clearly he has a lively time ahead.

The Banks

Beirut is the hub of a network of banking interests, and has about eighty functioning banks. Money flows into this entrepôt from all over the Middle East—and elsewhere—from insurance services, tourism, trade in general, and, above all, oil. The Kuwaitis, as noted above, found this city the most convenient nearby repository for their enormous streams of wealth derived from oil—as did the Saudis and such remarkable individuals as the ruling Sheik of Abu Dhabi on the Persian Gulf, whose income has been calculated at one million dollars a week. The Middle East is nothing if not a pool of oil, and Beirut has taken in a large share of the revenues. This is a capital which, above all, concentrates on what it considers to be the divine process of making money, and making money talk.

The banks have laws and regulations even more favorable to investors than Switzerland. Numbered accounts, which may or may not be secret, are at the disposal of almost anybody, but the Beirutis go the Swiss one better. In Switzerland a bank may be compelled by court order to divulge the name of a depositor; here in Beirut this is not the case—an account is secret and inviolable, no matter what. Beirut banks are, in general, well equipped, well run, and prosperous, although there are so many nowadays that they touch elbows. Normally, they can afford to pay interest even on checking accounts.

Even so, part of the banking structure is flimsy, and banks have been popping here lately like Mexican beans on a hot plate. I noticed on arrival at the airport that the bank which normally cashes traveler's checks for tourists, et cetera, was dark with a sign on its window, "CLOSED FOR INSOLVENCY." At least five banks failed or had to be taken over by the Lebanese Government in the summer of 1968, making a total of eight since October, 1966. Among them were the Banque al Alhi, one of the oldest and best-known national banks in the country; the International Arab Bank, an important institution (although new); and three small banks catering to private depositors, the Lombard Bank of Lebanon, the Professional Trust Bank, and the Lebanon and Middle East Bank.

The government is extremely attentive to the banking picture, and wary about it as well. It has no wish to kill the goose that has been laying so many comforting golden eggs. A man named Pierre Edde, the chairman of the Lebanese Banking Association as well as president of a bank of his own which combines Lebanese and Saudi capital, was appointed to be Minister of Finance in July, 1968, with orders to clean up. Three major reasons were given for the economic crisis that seemed to be afflicting Lebanon —the aftermath of the Arab-Israeli war in 1967, the international disturbances in the gold market early in 1968, and continuing repercussions of the most dramatic failure in the history of Lebanese banking, that of the Intra Bank in 1966.[6]

This episode deserves a word. An ambitious young man named Yusuf Beidas (sometimes spelled "Bedas"), born in Palestine of Russian Orthodox parents, emigrated to Beirut after the 1948 war. Penniless, a refugee, he worked as a bootblack, then became a money-changer. Beidas was fantastically clever about money. He became rich almost overnight, knew how to give the right people the right favors, and by 1951, when he was still only thirty-nine, he set up a bank of his own called Intra, short for International Trading Company. Within ten years Intra became the largest, most aggressive, and most prosperous bank in Lebanon, with Beidas himself heading the most extensive commercial empire in the Middle East.

Some of his feats were remarkable. He knew that primitive Arab sheiks, who made stupendous sums from oil, had no experience of banking and sat on their money, refusing to put it into any bank at all. He cozened deposits out of them by giving them confidence—he would arrive at a tribesman's tent with pillows stuffed with cash, to prove that he had it, or, dealing with more sophisticated Arabs, would make a standing offer to cash a check

[6] *New York Times,* July 13, 1968.

for *any* amount, without limit, on sight. His Intra Bank throve. He not only tapped the Arab hoarders, but went into big business abroad; he bought real estate in Paris and New York, using deposit money, until he had assets totaling more than a billion dollars—on paper—and held a fifth of *all* Lebanese bank deposits.[7]

Nemesis was, of course, lying in wait right around the corner. Beidas borrowed money on short term, and then tied it up, out of overconfidence (he himself blames a personal conspiracy against him by other Lebanese bankers and government officials), in long-term investments which could not be liquidated quickly in a time of crisis. The crisis came. The world-wide credit squeeze began early in 1966, deposits shrank, and he was having to pay 9 percent interest for money he could only lend at 7. For a time he paid dividends out of capital. Meanwhile some of his big Arab clients had begun to travel, and discovered other banks in the world outside. The crash came on October 14, 1966. Beidas had to have overnight something like $230 million, and could only raise $30 million. His bank closed its doors; there were riots on the streets; the shock to the Lebanese financial system was such that *all* Beirut banks shut down for three days. No similar catastrophe had occurred in the international banking fraternity since the collapse of the Credit Anstalt in Vienna more than thirty years before.

Beidas fled to Brazil, then to Switzerland, where he still lives. The Lebanese Government does not press too hard for his extradition, probably because he knows too much. His dossiers are reputed to be like those of the secret police. After fifteen months the Intra Bank was put together again and refloated. Its four principal creditors were the governments of Lebanon, Kuwait, and the sheikdom of Qatar, plus, of all things, the United States Commodity Credit Corporation, with which Beidas had been involved in wheat purchases. Twenty-one percent of Intra's

[7] *Time*, October 23, 1964.

debt is, in fact, owed to the CCC, and an American official sits on Intra's new board to protect its claims.

Another fabulous figure in the world of Beirut finance, but of a different stamp, is Adrien Geday, also a Palestinian with nerves of iron. He began his career in Beirut as a peddler of taxi licenses, and is probably today the most powerful banker in the country. He is cooler, smoother, more conservative, than Beidas, but these two entrepreneurs, although rivals, worked together closely—when circumstances made this convenient—for many years. Geday owns the Beirut Casino, which he bought with funds supposedly borrowed from Beidas. His wife is Jewish, and he leads an unobtrusive life.

☼

I asked authorities about Beirut's economic affairs in general. The most interesting answer was that the third of the population which is engaged in trade and commerce generates two-thirds of the national income, but, conversely, that two-thirds of the people have only one-third of the total wealth to share. The gap between the rich urban class and the poverty-smitten rural areas is wide, and certainly the Lebanese Banking Association, comprising both Christians and Arabs, is the most powerful lobby in the country.

Food, Drink, and Fun

The standout attraction in Beirut is the Casino, which lies about half an hour by car from the Saint-Georges along the scintillating rim of shore. The blinking lights—like fireflies—are not strong enough to illuminate the maggoty detritus along the streets. The Casino itself, which is immaculate, is the largest in the world, and probably the most brilliant enterprise of its kind anywhere between Paris and Tokyo. There are chambers for roulette (the least expensive chips are the equivalent of 30 cents), *chemin de fer*, craps and blackjack, slot machines without end, a variety of bars and restaurants, and, above all, the show. This performance goes on every night, and lasts for almost two hours. It resembles the gaudy spectacles in Las Vegas or at the Lido in Paris, but is more old-fashioned and cleaned up—no bare nipples are permitted. One mad extravaganza follows another—a troika of three horses treading "snow," a girl swimmer in a diving bell, camels, a railway train spouting out live steam, a belly dance (of course), a pageant depicting a merchant of Baghdad visiting the royal court, a rainstorm, girls on moving platforms, girls on trapezes swaying out over the audience, girls suspended in hoops thirty feet high, girls, girls, girls—who are mostly, as a matter of fact, not as young or pretty as they might be.

Some of the Beirut restaurants have substantial reputations, but they do not amount to much. One is supposed to serve fifty-five different kinds of hors d'oeuvres, a specialty called *mezzah*. The cuisine varies between the French and the Oriental, and Beirut is a good city, as is natural, for stuffed peppers, shashliks, and kabobs. Cherries, tomatoes, and apricots are all succulent, and seem to be oversized; they reminded me of the old joke that grapefruit come so big in Texas that six make a dozen. The pistachio nuts are the size of almonds. The best and most luxurious restaurant in the city is the Saint-Georges grill, with its delectable marine view. Some Arabic words for foodstuffs are fun to try to learn, like *feliflat* for pepper, *shamman* for melon, *mawz* for banana. But salami is salami. One restaurant, the Flying Cocotte, has "very-in" (*sic*) specialties to mark each day of the week—spaghetti psychedelic, gratin Bonnie and Clyde, émincé de boeuf Carnaby Street, and Piccata Blow Up. On the menu of another, the Al-Ajami, I saw some of the quaintest misspellings of English I have ever come across except in a hotel in Ravenna, Italy—"Spiced Eggplan," "Ladies fingers-rice," "Sluffed lam meat," "Conserved lamp meat," "Brains-Salad," and "Sheep testacies." The national drink is, of course, arak, a form of Pernod, and the local beer is not too bad.

Streetwalking is legal, and across the street from the gold market (the juxtaposition is probably accidental) stands a flourishing *quartier réservé,* one among several in the city. The most celebrated of the local houses of ill fame, that of Madame Afaf, had to close down, however, because of a recent scandal. A politician of the most eminent rank and noble antecedents visited this establishment one evening, and discovered his own daughter there as an inmate. He dropped dead of shock.

Up the Hill

It was 90 in the shade, and the air resembled a wet sponge. Beirut is a tropical port in summer. We rose out of the susurrous yelp of the local traffic and ascended the hills swiftly; our chauffeur reminded me of the man in an old Max Beerbohm story who drove "as if he were threading a needle from afar." Almost at once, up fifteen hundred feet, we could see the city and ocean spread out before us; this might be Positano, a Greek island, or some equally entrancing hem of Mediterranean shore. Our destination was Baalbek, the ancient Heliopolis (City of the Sun).

Our course ran parallel at first with the line of the German-built (1909) rack-and-pinion railway traversing the ninety-one miles between Beirut and Damascus. To accomplish this, even today, takes eleven hours. The train carries nothing but freight these days. The 1912 Baedeker says: "Luggage must be at the station not less than ¼ hr. before the departure of the train. Travellers are strongly recommended to have the exact fare in readiness. Ladies should travel first class, but gentlemen may use the second class without fear, as there are also third class carriages on the trains."

In a comfortable automobile my companion and I reached the neighborhood of Mount Lebanon[8] through a pass called Dahr

[8] If one wants to be pedantic there are three usages of the term "Mount Lebanon." (1) It can mean the Lebanon range itself, the backdrop to Beirut, in contrast to the Anti-Lebanon range across the valley. (2) The craggy hills to the south of Beirut, which were not under the control of the capital in Turkish days but which were ruled by the Pasha of Sidon. (3) A particular short range just to the south of Beirut, known correctly as the Djebel Lubnan. Mount Hermon, the best-known peak in the country, near the Israeli-Syrian border, is not visible from Beirut.

el Baidar (White Back), and floated, so it seemed, over the shimmering hazy trough of the Bekaa Valley, one of the prime sights on earth. It lies between one Lebanese range and another, and is cultivated in long fertile parallelograms under the striped snow; geologically it is the last extension into Asia of the Rift Valley, which bisects Kenya a thousand miles away and cuts steadily northeastward through Ethiopia toward the Middle East. A handful of rich Beiruti merchants own the Bekaa. We stopped at the village of Chtoura, nicknamed Honeymoon Town, for obvious reasons, engagingly located in the lower mountains.

Here we drank some of the local wine, made by French Jesuits, at a pleasant small hotel with a garden bursting into bloom. The wine cost 4.75 Lebanese pounds ($1.56) a gallon. Proceeding, we saw chicken farms (established with the help of the American Point Four program), a few Bedouin, a baby camel, and gypsies moving by truck. The Bedouin also use Diesel trucks nowadays to circulate from pasture to pasture, tending their sheep. Indeed, the modern world has poked its way into the most ancient crevasses here. We passed through Zahlah, the third biggest town in the country, still in the valley, and moved on to the great Roman ruin of Baalbek, across the plain, on rising rocky ground.

Even the small temple here is bigger than the Parthenon. These Herculean columns are the stoutest ever built by the Romans, and, to this day, nobody can be sure exactly how the massive drums of marble which compose them were lifted out and put into place one on top of the other, to say nothing of their friezes. This was a whole city, not merely a temple, in Roman days—the capital of the local colony under Augustus. Behind the Roman façade lurks a hint of dark Carthaginian rites and secret processes. The temples were hurled to the ground and largely broken up by an earthquake in the 1750's. A summer music festival is put on in the ruins today, enchantingly illuminated by night, which is sometimes distinguished by such participants as Herbert van Karajan.

Baalbek is without doubt a noble ruin, but sad as well—almost sinister. This was the third time I had seen it, and I felt that it had become grossly vulgarized. Voracious guides and touts grab at the visitor, who is also surrounded by coveys of wild children some of whose lips were covered with chancres and with eyes eaten out by disease.

Beirut International

I have several times indicated Lebanese tolerance toward the Jews, but this does not mean softness to the Israelis. Lebanon and Israel are still technically in a state of war, and the frontier is closed. To go from Jerusalem to Beirut or vice versa necessitates a roundabout detour via Cyprus, which is a nuisance, even though Cyprus itself, where Venus rose from the sea, is pleasant to visit. This being said, it should be added that the Lebanese are much less inflammatory about Israel than the United Arab Republic (Egypt), Syria, or Jordan. Another point is that the country holds about 106,000 Palestinian refugees from Israel.

Lebanon has its own unique position in the Arab world, if only because it is half-Christian. The Middle Eastern nations are, to draw lines roughly, divided between the Arab "socialist" (by Arab definition) countries and the monarchies. In the former category are Iraq, Syria, Egypt, the Sudan—with a question mark— and Algeria. In the latter are Morocco, Libya, Jordan, Saudi Arabia, Kuwait, the minuscule ratty states near Aden and on the Persian gulf, Yemen, and, going further afield, Iran. Tunisia, like Lebanon, is a special case. By and large the United States gets along better with the conservative monarchist group than with the Arab nationalist states, which to one degree or other are associated with the Soviet Union, particularly Syria. Lebanon, although a close American friend, has, however, perfectly normal relations with the Russians. The Soviet Embassy, located on the site of an old caravanserai for Russian pilgrims, is conspicuous. Another factor to mention is that, as everybody knows, Soviet *naval* power has lately become heavily reinforced in the whole region of the Eastern Mediterranean, centering on Egypt, and this is felt locally.

Finally, in regard to Beirut and the Lebanon there should be a word about the boycott. It is impossible in this city and country —so passionately devoted to commercial enterprise, including imports—to buy Coca-Cola, a Ford, or anything made by RCA. This is because the Arab Boycott Bureau in Damascus has put these companies on the blacklist for Arab importers or consumers, for the reason that they do business in Israel. It might well be asked why Coca-Cola and so on do not withdraw from Israel, a market limited to two and a half million people, in order to gain eighty million or more Moslem customers. The Beiruti answer is that this is caused by pressure from American Jewry, which may—or may not—be the case.

Superfinally, the question might be asked, "Does Lebanon exist?" What is the cement holding this patchwork of minorities together? In a sense the country is a miniature version of the old Austro-Hungarian Empire: it fills a vacuum, and it would be necessary to invent it if it did not exist. Lebanon has a larger experience of liberty and a longer practice of representative government than any Middle Eastern country. One able Beiruti intellectual, Dr. George Hakim, a former Foreign Minister and Lebanese representative to the United Nations, now associated with the American University of Beirut, put it to me this way: "Our cement is triple—the inheritance of the French, desire to be a sanctuary, and love of freedom." He did not explain why Beirut itself, for all the supernal beauty of its *mise-en-scène*, is such a mess.

15. A FOOTNOTE ON AMMAN

This is one of the least-known capitals in the world, one of the most inaccessible, and one of the pleasantest. The Jordanians are the most agreeable Arabs I know, and their principal city, remote and humdrum as it is, reflects this *gemütlich* quality. Correctly Amman is spelled 'Ammân, and officially the country is called the Hashemite Kingdom of Jordan, or Al Mamlakah Al Urdunniyah Al Hashimiyah. Arabic is a language that delights in complications, as may be gathered from the fact that there are a thousand different ways of saying the words "camel" or "camellike." The language has no cognates. For a beginner to learn two hundred words in six months is par for the course.

Jordan—and Amman—are full of curiosities. The country was invented by Winston Churchill, then the British Minister for Colonies, shortly after World War I, out of the desert sands, partly to protect the new oil pipe lines being laid out. It was named Trans-Jordan, and its first ruler, Emir (later King) Abdullah, a roly-poly shrewd, and firm-minded man, had been one of Colonel T. E. Lawrence's principal raiders and a leader of the Arab revolt. Once, it has been recounted, he went to a Western movie in Jerusalem, and saw pretty girls unveiled; he rolled with astonished laughter, saying, "God is great!" I met him once in Amman in the early days. There in the entrance hall of his palace was an extraordinary object, a very large concave-convex mirror of the kind seen in amusement parks, which produced madly distorted images of those who looked into it, as all visitors to the palace had to do, hauled there by Abdullah himself.

The two most interesting things about Amman today are (a) the person of the present King, Hussein, and (b) the refugee problem. But before going into these we should have a word of background.

Jordan is a vestibule-sized country shaped like an axhead, and lies immediately east of Israel on terrain composed largely of rock, lava, and sand. Its territory included the West Bank of the River Jordan, together with East Jerusalem, as we know from preceding chapters, until the Israeli occupation after the Six-Day War in June, 1967. Today the river itself is the frontier. The country's chief product, next to trouble, is phosphate. One unusual—perhaps unique—fact about it is that, with a population around two million, it contains not a single Jew.

The Six-Day War dealt Jordan a desperately serious blow, if only because—temporarily at least—the West Bank, including the Arab sector of Jerusalem, was lost. This is as if New York State should have lost Manhattan. As one authority puts it, the West Bank contained half the population of the country, more than a third of its arable land, and at least half of its income in foreign currency. Jerusalem alone (the Jordanian sector) had been worth $35 million a year in tourist traffic.

The river frontier lies only about forty-five minutes by car from Amman, and is extremely agitated. Its principal point of crossing, the Allenby Bridge (called the Hussein Bridge by the Jordanians), spans the Jordan north of the Dead Sea, the lowest spot on earth, and is a focal point for serious disturbances, although merchants, traders, and so on cross it freely by day. Along the Jordanian side, called "the Valley," which is irrigated to an extent, are orchards, vegetable farms, and so on, vital to the economy. Here the Israelis make provocative raids from time to time, including one full-scale punitive attack on a village called Karameh in the spring of 1968, in which fifteen thousand troops were involved. The Israeli motive is to destroy the Jordan irrigation system and water supply as well as play havoc with the countryside.

An attempt to counter this by the Arabs has been made, as mentioned in a previous chapter, by the establishment of three

different commando groups—Al Fatah with its military wing
known as Al Assifa, the Front for the Liberation of Palestine,
and the Palestine Liberation Army. The first, which is by far the
most important, is supposed to have anywhere between six and
fifteen thousand men under arms; its recruits are trained in
Egypt, Syria, and Algeria.[1] The activity of these organizations
has become a vivid and controversial issue in Jordanian politics.
King Hussein cannot easily disavow them, but he does not particularly enjoy their prominence, because the Palestinians are a
large and distinctly unstable portion of the population and they
do not hold the King in high regard, a sentiment which is heart-
ily reciprocated. What the King wants above all is a *settlement*
—on acceptable terms—with Israel. But he has to tolerate the
terrorists, which means that he stands on a razor's edge.

The conduct of the border warfare continues to be merciless.
If a Jordanian kills an Israeli, the Israelis will kill—if they can—a
dozen Jordanians in retaliation, swooping in silent formation
across the frontier at night and cutting up a village. The equa-
tion, as the Israelis see it, is not an eye for an eye, but ten eyes
for one.

The history of this small country, panting on the desert, goes
far back. Originally it consisted of such ancient Biblical "kingdoms" as Edom, Moab, and Gilead.[2] South of Amman rose the
city-state of Petra, "the rose-red city half as old as time," the
ruins of which are still the country's chief tourist attraction, now
that East Jerusalem is gone. Then, as in Israel, came continuous
waves of foreign invasion and occupation over centuries—by
Egyptians, Assyrians, Greeks, Romans, Crusaders, Turks, and, at
the end, the British. The Turkish administrative center for the
area was neither Jerusalem nor Amman, but Nablus, now in
Israel. After World War I and the victory of the British, French,
and Arabs over the Turks in this area, Trans-Jordan was con-

---

[1] Dana Adams Schmidt in the *New York Times*, September 24, 1968, also
Thomas F. Brady, *ibid.*, April 4, 1968.
[2] Here I am following a document issued by a neutral embassy in Amman.

sidered to be part of Palestine. Then, however, London took the line that its pledge to help build a Jewish National Home in Palestine did not apply to Trans-Jordan (an effort to whittle down Zionism and give comfort to the Arabs), although it remained part of the Palestine mandate. This made for an anomalous situation, bitterly resented by many Zionists. In those days the imperialist powers cut up territories (and documents) like butcher paper.

In any case, something called "Occupied Enemy Territory East" was set up, with its capital at Damascus, now the capital of Syria. An Arab prince, Faisal, who had fought brilliantly with Lawrence to help liberate the Arabs, was established there as king. But the French kicked Faisal out in July, 1930. The idea arose to appoint Faisal to Trans-Jordan, which, still part of the Palestine mandate, was being ruled by the British High Commissioner in Jerusalem. This suggestion did not materialize, and Faisal became King of Iraq instead. The British then called a number of Jordanian sheiks together, and decided to make Abdullah, Faisal's younger brother, the Emir (or Prince) of an "autonomous" Trans-Jordan under British control—a purely makeshift and arbitrary creation. Mr. Churchill, as mentioned above, was the decisive factor in this maneuver. Abdullah consented to take the throne, and the emirate was granted complete independence in 1946, with Abdullah raised in rank from Emir to King. The country was named "the Hashemite Kingdom of Trans-Jordan" because Abdullah is a member of the Hashemite family, which claims direct descent from Mohammed over thirty-seven generations.[3] Hashemite is an Anglicized adjective relating to Mohammed's clan.

Abdullah led Trans-Jordan into the war made by the Arab states against Palestine in 1948, and, although the Arabs got a whacking, he was able later to hang on to the West Bank, including East Jerusalem. As one historian put it, "he unilaterally annexed Arab Palestine to Jordan," and the fact that he did this

[3] "Hussein Approaches 'A Point of No Return,'" by Curtis Bill Pepper, *New York Times Magazine*, April 7, 1968.

on his own without juridicial sanction from anybody is the principal legal reason why the Israelis claim the West Bank to be their own today. In April, 1949, Abdullah changed the name of the country officially from Trans-Jordan to Jordan, to mark the new extent of his realm. But he was assassinated two years later while praying in the Al-Aqsa Mosque in Arab Jerusalem.

His successor was his son, Talal I. This young man was, however, unfortunately insane, and he abdicated in 1952. Hussein, the present King, who was the son of Talal, ascended the throne in 1953, after a period of regency.

Amman, the Capital

Amman is one more capital built on seven hills, but here the hills are substantial, not buttons as in Rome or virtually invisible as in Moscow. To circle or climb them can be laborious. The city, which lies at an altitude of about 2,500 feet, has a continental climate, but it can be very hot in summer. Except in the older parts, which lie low and resemble an Oriental bazaar, with lines of squalid traders' shops, most buildings are hewn out of limestone from the surrounding hills; their windows look like holes in rock. The prevailing color is ocher, almost orange. Water is plentiful, since the city rises from an oasis. Altogether the views provided by Amman, particularly at dusk, are as breathtaking as in Jerusalem, and yet the city as a whole does not quite have the character of Jerusalem, Damascus, or even Aleppo.

There has been an astounding population boom, largely because of the influx of refugees. When I saw Amman first thirty years ago, it was no more than a shadowy Arab village; today it holds more than 450,000 people (twice the population of Jerusalem) and is expanding rapidly; new houses sprout on the great hills like stone mushrooms. A good many have TV antennae. Amman has one TV station, comparatively new, with one channel—"Jordan Television." Plans are going forward to put up a station for satellite communication; perhaps this is a demonstration of

Arab nationalist rivalry, because Kuwait, Saudi Arabia, and the Lebanon are also installing this new form of apparatus, and Jordan does not want to be left behind.

Amman is the ancient Philadelphia, the name given to the site by Ptolemy II Philadelphus, King of Egypt (285-246 B.C.), who ruled here for a brief period after the original city had been destroyed by the Persians. Before this it was known as Rabbath Ammon, the capital of the Ammonites, who were, if the reader will remember his Bible, the descendants of Lot. Under Roman rule Philadelphia flourished, and the chief historical and archaeological sight today is the ruin of a handsome Roman theater, which had a seating capacity of six thousand. Theatrical performances are still put on here occasionally. Other antiquities are to be seen as well, like a Temple of Hercules and a substantial forum. After the Romans came the Arabs, and the city crumbled; it became a minor provincial town, and declined as the Arab empire did. Under the Turks it was nothing but a neglected village. Then World War I brought it back into being.

Amman became important at that time partly because it was a station on the celebrated Hejaz railway, which runs from Damascus through Deraa (where Colonel Lawrence was captured and debauched by a Turkish pasha), Amman, and Ma'an. I have ridden on this railway, and on one occasion I was stranded for several days in Deraa, a filthy town; I would not like to repeat the experience. So when I saw the rickety station here in Amman and was told that it was still a principal stop on the Hejaz railway (which in theory goes all the way to Mecca and Medina), I could not suppress a slight *frisson*. Nowadays the railway is not much used because a new double-lane asphalt highway, built by the Saudis and British, connects Amman with the holy cities. Now Amman also has a small airport, with modest signs saying "THANK YOU, COME AGAIN." A large, intricately designed Oriental rug hangs on the wall of the reception room, but this typical Middle Eastern artifact was, I learned later, made in Belgium.

Once you get away from the hills Amman is laid out straight, and, outside the bazaar area, is spotlessly clean. The proudest modern sight the city offers is its new sewage plant. One street, the name of which can be transliterated from the Arabic as either Mahata or Mohatah, is, like a street in Kansas, straight as a billiard cue and seemingly unending. I think it is the longest straight street I have ever seen outside the United States. One of its principal ornaments is the police station, a tall pale-blue building made of slate, with narrow windows, which citizens ironically call "Abu Rassul's Hilton." Abu Rassul was a recent chief of police, feared by almost everybody; recently he became an ambassador abroad.

Another substantial local sight is the athletic stadium being built more or less on the lines of Kenzo Tange's masterpiece in Tokyo, which will be named for Hussein; another is the university, founded in 1963 on the site of an agricultural exposition on the outskirts of the city; it has about 2,400 students, both men and women. The countryside nearby is moderately picturesque, dotted with quarries, walled farms, encampments of Circassians,[4] and ancient black watchtowers, which are made of stone dating from 800 B.C., the Iron Age.

One interesting street, or road, named Salt Street, leads toward Jerusalem, passing through the well-known old town of Salt. The Emir Abdullah, coming up from the Hejaz to rule Trans-Jordan, wanted to establish his capital in Salt instead of Amman, but the local people in Salt, a tough community, would not have him.

Amman has one old-established hotel, the Philadelphia, next to the Roman theater (which helps to give the city the flavor it has of continuous habitation over centuries), and a sparkling new establishment, the Intercontinental, which seems to me the best hotel I have ever encountered in the Middle East. It has gadgets

[4] The 1912 Baedeker has a tart little line, "The government has established a colony of Circassians here, unfortunately not to the advantage of the ruins [of Amman]." The Circassians were brought in by the Turks as a gendarmery. Their startling blond descendants are sometimes seen among the darker Semitic populace.

as contemporary as anything in New York, and the service rivals that in a really good hotel in London. One curiosity is that you may have jasmine in your coffee. I could not decide at first why the swimming pool, seen from the long comfortable deck which holds lunch and coffee tables, appeared to be so charming, and perhaps the reason is that it is not laid out on sand, as pools generally are in this part of the world, but on a bright green lawn.

The public services of the city are, all things considered, well maintained. The buses have to circle the hills and, as a result, their routes seem to be unendingly roundabout, but the operation is efficient. "Service taxis" hold three or four passengers going to different destinations on a fixed route; the fare is about a nickel. The Mercedes is by all odds the favorite private car, and there are more automobiles per capita than in Jerusalem. Education and public health are responsibilities of the national government, not the city. The municipality depends mostly on a kind of real estate tax (like New York) for its income, and a seven-year Development Plan is in progress.

The most interesting official I met in Amman was Ahmad Fawzi, the Minister of Public Works, who previously served four years as Deputy Mayor of the city, and then became Minister of State for the Prime Minister's Affairs (a title which, so far as I know, is unique in the world). Mr. Fawzi, both able and sophisticated, a Westerner in his habits of action, was educated variously at the Universities of Virginia, California, and Wyoming. He said that one of the curses of public administration in the Middle East was that everything had to be done personally; it was impossible to deputize authority. He sketched some of his problems. Amman has very little crime, he said, which is the more remarkable in that many citizens bear arms and are addicted to blood feuds. Seeing Mr. Fawzi was interesting on another score aside from his talk, because of the Oriental paraphernalia attached—his large room in a cracked-plaster building was lined with benches, on which sat Arab supplicants, petitioners, and retainers, in decrepit rags. One by one visitors with appointments were brought to Mr. Fawzi's desk, but the supplicants never moved, waiting stolidly in his own office, and resembling rag dolls pulled out of a duffel bag.

The people of Amman and its hinterland are divided (refugees excepted) roughly into two classes—the Bedouin, who are pastoral nomads, and the old Palestinians. These latter, whose wealth was based on land on the West Bank, are the elite—many merchants made large fortunes after the war, when inflation brought the price of a tire, as an example, to $1,000. They are commission dealers, money-lenders, traders—also intellectuals. There is, however, a considerable chasm in Jordan between rich and poor, as in Kuwait. The poor are probably better taken care of here than in any Arab country, although no social security exists. Taking the place of this are family relationships, charity from Christian missionaries, and donations from the *Waqf*, a Moslem religious foundation. People really poor are miserable—on the very edge of destitution. Forty percent of the population is under the age of sixteen—this is a young society. A good many young people seem to be aimless and discontented, with nothing to do beyond holding down a precarious job, if they have one. A visitor feels little of the sting, energy, and sense of emergency that distinguish Jerusalem. Many young people brighter than the ordinary or otherwise well favored do, however, manage to escape abroad to complete their studies or become teachers and technicians. One estimate is that there are no fewer than 25,000 Jordanian university students in colleges all over the world, many in the United States. The young official who met me at the airport had just returned from the University of Oklahoma, where his "thesis," as he described it, had been a study of the work of a contemporary American journalist. It was pleasant to be met by this particular young man.

The biggest export of the country is, I heard, talent. The Jordanians (like Palestinian Arabs in older days) branch out everywhere as teachers, technicians, civil servants, businessmen.

His Majesty the King

The monarch, King Hussein, is an agreeable young man, discreet, earnest, soft of voice, well mannered—perhaps somewhat ineffective as well, but not in the physical sense. Indeed, the contrary is true. Hussein, who is in his early thirties, has unlimited courage, and goes in for the most muscular sports—water skiing, scuba diving, and "parachuting" on a towline from a speedboat, which may lift him two hundred feet above the water. He likes automobile racing as well, and flies his own plane, a Caravelle, to various European capitals. Once he took the controls on a commercial airliner to New York.

His intellectual pursuits are more limited. He cares little for books. Impetuous, he makes up his mind on issues at the first or last moment, often without consulting his staff; he doesn't take advice easily, and is called "unpredictable." But he has a considerable sense of mission and interest in the complicated history of his family and its roots. His mode of life is informal. He does not live in the Basman Palace in town, where he works, but in a small villa out in the hills, where he likes to help in the cooking —even in setting places at the dinner table when guests are coming. His personal life is altogether quiet.

Hussein was educated at Sandhurst. He has been married twice. His first wife, a member of the Hashemite family and a distant cousin, was Princess Dina Abdel Hamid; they were married in 1955, and divorced two years later. Princess Dina was an intellectual, Cambridge-educated, who had a job teaching school in Egypt. They had a daughter, Princess Alia, now aged twelve, for whom the Jordanian Royal Airline is named. Next (in 1961) Hussein married an English girl, Toni Avril Gardiner, whose father was an officer in the British military mission in Amman. Hussein met Miss Gardiner at a party; nowadays she is known as the Princess Muna. Four children have been born of this marriage, including twin girls last year. The crown prince and heir apparent is Hussein's youngest brother, by name Hassan Bin Talal. He married recently a Pakistani girl, the daughter of a distinguished diplomat whose wife was also a professional Pakistani diplomat. A commoner, she became Princess Sarwat. Her husband met her when they were both students at Oxford.

King Hussein gave me an audience at Basman Palace, and, as is often the case in an interview with a royal personage, the most interesting feature was the quality and amount of ritual attached. Here the procedures were modest. Surrounding the palace are thickets of oleander and jacaranda, with its delicate violet color. There were several varieties of guard—first, Arab warriors on the steps outside, wearing white-and-red headdress, who stamped their feet as a salute. Inside were palace guards with red berets, some in khaki shirts, some in pale green, and then a group of Circassians—who looked like giants, in their tall black boots, with silver bullet pouches crossing their blue shirts, and wearing a short black tarboosh. I was passed from room to room, meeting members of the royal establishment one by one, each higher in rank than the man preceding. This process took about an hour and a half. In the last anteroom I was photographed by a cameraman who leaped in unannounced and took blinding shots in the manner of official photographers in presidential palaces in South America.

Hussein received me alone—no secretaries, no officials. I had not known that he was short in stature—only about five feet four. He was in civilian clothes, with a regimental tie, and wore a large sportsman's wristwatch. His eyes, set widely apart, are quite small, but they glisten brightly. The King seemed shy, a bit out of focus. He talked well, but was guarded in what he said. "People say that we are reactionary, but no—we are merely conservative." "Refugees should be an international problem." I tried to draw him out on the characteristics of Amman as a city, but he did not respond.

Hussein has not had an easy time in his years on the throne. There have been several attempts to assassinate him, and he has not forgotten that both his grandfather, Abdullah, and his cousin, Faisal of Iraq, were murdered by political assassins.[5] In fact, he was standing directly next to Abdullah when the latter was shot down—a traumatic experience. He has had to put down several insurrectionary movements. There came a severe crisis when he discharged Glubb Pasha, the obstinate colonial-minded British officer who had served Jordan for a generation. Hussein has had troubles with Nasser, with Iraq, and with his hereditary enemies the Saudis. Seventeen thousand Iraqi troops entered Jordan during the Suez crisis in 1956, presumably to help defend the country if the Israelis moved in, and have never left. Above all, he is first on the firing line vis-à-vis the Israelis, and has an appalling refugee problem on his hands. Mainly Hussein has managed to hold on to power through the support of the Bedouin. He himself is a Bedu, and he bases himself on a Bedouin tribal foundation.

Suffocation by Refugees

Here in Amman and on its outskirts lies one of the largest concentrations of refugees in the world, and its miseries are boundless. A friend drove me past the camp called Wahadat, on the slopes of Eshrafiya, the highest of the city's seven hills. Here live between fifty and sixty thousand Palestinian Arab refugees who date from 1948, after the Israeli war of independence. Twenty years later, they still live in wretched mud huts, jobless for the most part, devoid of skills, energy, or ambition, in circumstances of the utmost squalor. People sleep twelve in a room, and are a trying drain on society. Next we went to the heights of Swaileh, a Circassian settlement, and looked down on the refugee camp called Baqua, with 32,000 inhabitants; these are also Palestinian Arabs but of a different decade—they came here after the June war in 1967. With the exception of a few in corrugated-iron shacks, they live in tents. Rainfall can be severe in Amman, and

[5] Pepper, *op. cit.*

their misery may be imagined. But even in the worst *bidonvilles* the children seemed clean and well cared for. Some Baqua residents are known as Taani Marra, or "two-time" refugees; they have been displaced twice, first from one part of Palestine to another after the 1948 war, second from Israel across the river into Jordan because of the dislocations made by the June war in 1967. Five or six persons live in a tent. Families are irremediably split, because the members did not necessarily set out for Jordan at the same time. It is difficult in the extreme to give this latter-day type of refugee an occupation or resettlement. They still have hopes of returning to their former homes across the river and bitterly resent being given permanent quarters. Yet they cannot be abandoned.

Accurate figures are difficult to assemble, but good estimates have it that about 350,000 Palestinian Arab refugees now located in Jordanian camps entered Jordan after the 1948 war, and about 250,000 since the Six-Day War in 1967—a total of 600,000. If refugees not in the camps (some are taken in by families or friends) are included, the figure is much higher, say 880,000. Yet the total population of Jordan is only 2,130,800, which means that roughly one out of every two and a half persons is a refugee. Scarcely any small country can be expected to support such a harrowing burden indefinitely.

In plain fact, most of the refugees are kept alive by UNRWA, the United Nations Relief and Works Organization, but the Jordanian Government contributes what it can. Completely destitute families, known as "soulless ones," get a dole of wheat, free medical services, and free schooling for their children. The problems are humanitarian and psychological as well as economic and in the realm of public health. It is extremely difficult for a refugee to become a nonrefugee, and resume a normal status in society, chiefly because of the lack of jobs. But a few become self-supporting shepherds in the countryside near Amman, and some young men actually break all the way out and go abroad to seek employment in another country, like Kuwait. But the hard core doesn't stir. One difficulty is that the Arab world as a whole has never been able to join forces to deal with the refugee problem on an over-all or systematic basis. A thing to be said for Jordan is that the Jordanian Government gives the refugees citizenship. Syria and the Lebanon do not, for complicated internal reasons of their own (for instance, Lebanon does not want to enlarge the Moslem component of its population), and this is an anguishing matter in those countries because a person cannot get a job without a work permit, and these are not given to noncitizens.

A new wave of refugees came from the *east* bank of the Jordan into Amman and its neighborhood after the Israeli attack in force on Karameh in 1968, and as a result of Israeli raids on border communities in the Valley. Jordanians lost their farms and fled inland—there was no other place to go. To this miserable refugee problem and all its ramifications there seems no end.

Is Jordan Viable?

Surprisingly enough, the economy of the country is in fairly good shape. But is Jordan viable? Can it live without Jerusalem and the West Bank? Can Hussein control the vast number of incoming Palestinians, who are bound sooner or later to dilute his Bedouin support? Does the country possess a true national identity? The name "Jordan" (or even "Trans-Jordan") does not appear in T. E. Lawrence's *Seven Pillars of Wisdom*, although this is the territory where his activity was concentrated. Newspapers in Kuwait today, also Syria, almost never use the term "Jordanian." Nationalist roots are shallow here; Jordan was no more than a *tribal* area until the most recent times. Children in Amman entering the schools do not know the Jordanian national anthem, and a few youngsters, so I was told, do not even recognize their own flag.

The gold cover is high, trade is brisk, and, in spite of the Six-

Day War, which cost the country so much territory, the community has a moderately prosperous look. But a good deal of this is artificially stimulated. In plain fact, Jordan has been a country in forfeit ever since its creation—supported from outside. The British gave it an outright dole, called "budget support," from the formation of the emirate in the 1920's until 1957, when it was withdrawn following the Suez war and the British decision to curtail its overseas expenses. The United States stepped in promptly to replace the British in providing "budget support" until 1968. This was a kind of reward to Jordan for not helping Egypt in the Suez affair. Our direct budgetary help was, however, terminated in 1968 because of a feeling in Washington that the three oil-rich Arab countries—Libya, Kuwait, and Saudi Arabia—should assume the burden. Nevertheless American assistance to Jordan continues to be substantial, in such forms as economic and technical aid, the Food for Peace program, disaster relief, and various loans. In fact, the American contribution to Jordan since 1951 has totaled more than $540 million, not an insignificant sum. We also provide a Military Assistance program vital to the country's security. American military aircraft flew into Jordan, as a matter of fact, early in 1968—a substantial "airlift" to protect Amman when it seemed that there might be an Israeli invasion after the Karameh episode.[6] United States policy is firmly pro-Israel for the most part, but we want to keep on good terms with the Jordanians as well.

There are two reasons why the Western powers show solicitude over Jordan. First, the fear that the country might disintegrate, which could cause unending headaches in the Middle East. Second, fear of Soviet influence. Nobody in the West wants Jordan to become another Egypt. A number of Israeli leaders (but not all) take a similar view and favor a comparatively soft line on Jordan for the same reason. It is better for them to have Hussein in Amman than Brezhnev and Kosygin. Moreover, Jor-

[6] Joe Alex Morris, Jr., in the *Herald Tribune*, International edition, Paris, June 1, 1968.

dan is an obstacle to Arab cohesion and unity, and thus serves several Western interests.

✿

Now we take a long vault and cross half the world to visit a metropolis of a totally different order—Tokyo.

16. ARIGATŌ (THANK YOU), TOKYO

Tokoyo, the largest city in the world by population, is two-faced, double-sided, pig and porcelain. Its dominant note is commerce —money-making, money-spending—but its fierce and incessant financial activity and perfervid concern with the material elements of society are cloaked by a good deal of Japanese charm and hocus-pocus. This is Wall Street—or the busiest imaginable commercial community—in a kimono.

Tokyo is an exceptionally closed city as well, which makes the obvious stresses—between ancient custom and modern renovating influence, between "East" and "West," between the pursuit of fortune and powerful spiritual and semi-spiritual elements— harder to pluck and identify. There is always something behind everything; the city is a series of layers, or façades. Life goes by indirection. A rich man does not flaunt himself, but hides. The more prominent a citizen is, the more he seeks vigilantly to keep *out* of the public eye. The story is told of a photographer who came to Tokyo to take pictures of "society" for a fashionable American magazine; he couldn't find any society to photograph! Most of the patriarchs live totally unostentatious private lives. Their beautiful houses, sometimes situated on the tawdriest alleys, lie hidden behind stout walls. Tokyo faces in, not out.

Tokyo would be, indeed, just another very large, thriving, animated city, like dozens all over the world, if it were not for its essential *Japanese* qualities, which are what make it unique and, to my mind, so enthralling. Women still carry babies on their backs, even if they have graduated from kimonos to Western dress—and a modern-clad mother with a papoose is a strange sight indeed. But the women *walk* as if they were in kimonos still—with a peculiarly short, shuffling gait. Next to a modern office building you may find a Japanese inn fashioned exquisitely of wood with sliding paper doors. Even the most contemporary buildings have an exotic touch, like the new structure occupied by the Tokyo *Mainichi* and the *Reader's Digest;* the former is one of the three great metropolitan newspapers which have tremendous circulations—ranging up to 8.3 million a day. Seen from a distance, this monument, the largest office building in Asia, resembles the kind of flat cartridge—enormously magnified—that goes into an automatic camera, since it is flanked by two tall round towers shaped like spools. Lunching here one day I saw another example of Tokyo's startling contrasts. The Venetian blinds work by a push button, an innovation new to me. Open them and you see below the centuries-old stunted trees in the imperial courtyard. And you *do* look down on them and other appurtenances of the Emperor. Before the war this would have been impossible; to look *down* on anything having to do with the imperial family was a sacrilege.

One morning I stood by a window on the sixth floor of the Hotel Okura, one of the best hotels in the world, handsome in pressed brick and glass, to try to grasp the scene, which is higgledy-piggledy. Here in the very center of a city of ten million people are neat little gardens with trickling waterways, greenery everywhere, patrician dwellings, mottled back yards, flat roofs, factories, industrial trash, vegetable patches, tiny wooden hovels with the laundry hanging out, and lines of gingko trees. These, which are the particular symbol of the hotel, are widely planted in Tokyo because they are supposed to be a firebreak. The city —with reason—has an almost neurotic fear of fire.

The Okura is both cosmopolitan and intensely Japanese. I admired the sweeping, clean lines of the lobby with its chains of lanterns, the lack of clutter in the rooms with their two-deck ceilings, the pretty kimono-clad girls running the elevators, shoe-horns a yard long, shower heads shaped like telephone receivers, and the imperturbable waiter who, I am sure, if we had asked for a grilled pterodactyl for breakfast, would have answered "Medium or clisp?" without batting an almond eye. I noticed, too, that the radio set artfully into the bed table carried not merely the BBC and the Voice of America but Air India, Moscow, and Peking.

Street scenes are both drab and colorful. Most shops in the lanes and alleys off the Ginza-Dori, the main thoroughfare, and practically everywhere else in the huge metropolis, are windowless stalls like those in an oriental bazaar. Some have open boxes of Kleenex at the disposal of the shopper, since it seems that practically all Edoites, as citizens of Tokyo are often called, have colds most of the time. Most cashiers, even in the big stores, still use an abacus rather than a cash register—sometimes both. One symbol common to both East and West is the barber pole; there seem to be thousands of these, each flaunting its brightly moving color within the tube. Telephones come in three colors in Tokyo, red, pink, and black, and outside many shops, even on the meanest streets, bright red phones stand on posts in the open —pay phones for the convenience of customers.

Rickshas are seldom to be had nowadays, but fashionable geishas still use them—with shades drawn—and they may be seen parked sedately outside secluded restaurants. Always one has a sense of the inner tranquillity and inviolability of Tokyo no matter how agitated its contemporary movement or how Western its veneer. Like London, this is the capital of an island—a crowded island at that—and people protect their privacy by thickets of class, caste, and stylistic barriers.

But how beguiling most of it is! You can walk down a street and, on the open porches of an endless line of tiny houses, see the shoes left outside; Japanese never, as everybody knows, wear shoes indoors. There are flashes of green behind almost every gate, and water tinkles placidly over rocks artfully set in globules of garden. The public hullabaloo is terrific, but every house is a secret haven.

All this—not even mentioning such standard attractions as the traditional tea ceremony or fat men wrestling—helps to give Tokyo its mysterious dreamlike quality and peculiar charm. What could be more delicious than that wastebaskets are covered? A delicate small straw lid sits on the receptacle. What could be more exquisite than the design of even the simplest object—a bookmark or a teacup—or the artful style with which a package is wrapped? What would the Tokyo resident do without pebbles? They are part of the decor almost everywhere. If a house is too poor to have an actual garden, a tray (*bon-kei*) full of pebbles is set up in a room, and bits of green are stuck in this to simulate the reality of a tree or bush. What could be handsomer or more edifying than the way two elderly Japanese gentlemen of station, with their opaque faces, frock coats, and striped trousers, bow ceremoniously to one another when they meet?

Street scenes exist in other dimensions too. One intersection is nicknamed "Axis Corner," because it holds the offices of Japan Air Lines, Lufthansa, and Alitalia cheek to cheek.

I never heard an airplane overhead at the Okura; I never heard a bell, gong, or siren; I never heard a taxi howl; I never saw smut or grit. And I thought at first that this must certainly be one of the most remarkably tranquil of cities. I was wrong, because Tokyo boils with inner bluster in spite of its qualities of serenity and seclusion in some quarters. It gives an impression of tremendous confusion, overlying drive and energy, hard work, determination, and materialist effervescence—also of hysteria based partly on natural Japanese characteristics, partly on the desperate necessity to compete.

People, People, People

Tokyo, this complicated paradise and inferno all in one, is the biggest city in the world from the point of view of population, with 10,686,600 people under its metropolitan jurisdiction by the most recent estimate. New York and London have around eight million each. Tokyo is more populous than several sizable and important countries, like Sweden, or any but three American states. In area it is enormous too, covering 796 square miles as compared to New York's 319.8. It spreads out steadily if only because it does not grow high.

Tokyo is, indeed, eating its way remorselessly into the surrounding hinterland. The figures above refer to metropolitan Tokyo, measured by a 50-kilometer (roughly 31-mile) radius from the center of the city. But the entity known as "greater" Tokyo, with a 100-kilometer radius, is much bigger, with a total population of eighteen million as against thirteen million for New York in an equivalent space.

Satellite towns exist near Tokyo today even beyond the 100-kilometer radius, and add to the dense throngs of commuters fighting their way into the city day by day. Moreover, its population grows by some 300,000 a year, and the demographers estimate that "the capital region" will hold about 40 million people, a terrifying total, by 1985, with the figure for metropolitan Tokyo calculated at between 28 and 30 million.

Even today, some Tokyo statistics almost pass the believable. There are 32,940 police and 27,000 taxis (New York has 11,722 taxis). In a recent year 125,000 drunks were picked off the streets and arrested, a large proportion of whom were unconscious, according to a lively local guidebook.[1] Japanese, when they drink, drink hard. The city has no fewer than 97 universities, 93 newspapers, and 36 percent of the bank deposits of the nation. It produces about 20 percent of the total national income, and reflects vividly the economic growth rate of the country as a whole, almost 10 percent a year. A recent *Reader's Digest* survey shows that 44 percent of Tokyo citizens own their own homes, 69.3 percent carry life insurance, and 51 percent have telephones; 96 out of every 100 citizens have wristwatches, 77

[1] *Amazing Tokyo*, by Don Briggs, p. 93.

out of every 100 cameras. TV is practically universal, and approximately one out of every six homes has a car.

Prosperity is, in a word, the word. There may be a bust some day, but these are boom times now. One clear, if peculiar, reason for this is that Japan lost the war, and, like West Germany and Italy, the two other losers, leaped into a sustained postwar prosperity. By the terms of its peace treaty Japan renounced war and maintains only an extremely small (but expert) military establishment. The United States spends more than 50 percent of its budget on war, defense, and armament; Japan spends only a spoonful—between 1.6 and 1.9 percent. So there is plenty of money free to go around for other purposes, including industrial and commercial expansion. One index is that land in some parts of Tokyo has gone up 3,000 percent in value in the past few years; property near the Ginza can cost $3,000 a foot, and the city is probably the most expensive in the world, after Paris. A modest, conveniently located 3½-room apartment rents for the equivalent of $1,000 a month; a small rib roast of beef is $12, and oranges are $2 a dozen. One curiosity is that almost everything is paid for in cash; credit is tight, and personal checks unknown. Yet wages are scant by our standards, and nobody was ever able to explain to me how the very poor survive at all.

I have mentioned commuters. More than a million of these, depending on the season, enter the crunching maw of the city every day from the "dormitory" towns on the perimeter. There are plenty of desperately poor citizens in the outskirts, but few outright slums. No disgraceful shanty or squatter towns exist like those in Lima, Rio de Janeiro, or Santiago. One reason for this is the proliferation of danchi, low-cost housing units built by the government and otherwise; another is that few countryfolk migrating into Tokyo are destitute in the sense that villagers in South America are destitute. Most come from farms with a decent, if moderate, standard of living; nobody starves in Japan. Still another is that almost everybody entering Tokyo from the rural areas has an element of family already in the city—an uncle, brother, daughter.

So, generally speaking, the newcomer has a place to sleep until he finds a job and can fend for himself. One key to this is, of all things, the tatami mat. These straw mats, which measure about six feet by three, are the universal floor covering in Japan; a room is measured by the number of tatami mats it will hold, the figure for the average being four and a half. The Japanese do not sleep in beds. They pull a quilt out of a cupboard, and put this on the mat. When the family gets up in the morning, the quilts are rolled up and stowed away, whereupon the sleeping quarters become a living room. This helps to make the congestion bearable.

The satellite cities or dormitory towns are not, in the strict sense of the word, *sub*urban communities like their counterparts in the United States; they are just as urban as Tokyo itself, if smaller; most are full of factories. (An odd point is that there is no word for "community" in the Japanese language.) Some commuters come in from danchis forty miles distant, and the time given to commuting can be as much as four hours a day, which can make for exhaustion, frustration, and hysteria.

Why do farmers, peasants, workers, pour into Tokyo in such numbers? The answer is not, as might be expected, an advancing birth rate. On the contrary this is, in fact, very low—0.9 percent as compared to 3.5 percent or higher in Central America, as an example. Birth control is practiced by almost everybody, and abortions are legal. What does drive the streaming hordes into the metropolis is the variety and profusion of opportunity it offers—wider horizons, money, power, entertainment, and better education for the children. For Tokyo is a triple entity—Washington, New York, and the Groves of Academe all in one —the governmental, financial, and educational pivot of the nation.

A million commuters a day, on top of normal traffic within the city, make for agonizing problems. Tokyo was host to the Olympic Games, a big feather in its civic cap, in 1964, and this led to some badly needed face-lifting in the way of new elevated highways. But traffic is still convulsed. It is not, however, quite so bad as complaining Edoites say it is; New York is more congested, Rome and Paris more anarchic. But Tokyo had 77,472 traffic accidents in 1965, with 788 people killed and more than 55,000 injured, and the rate of 402.2 traffic deaths per 100,000 registered vehicles is the highest in the world. Tokyo police posts, of which there are about 1,300, have illuminated signs which announce the number of persons killed in accidents hour by hour. This is done by way of warning, not as an exercise in morbidity.

Traffic goes to the left, as in England. The pedestrian gets a break—sometimes. At some intersections he will find a kind of umbrella stand filled with yellow flags; crossing the street, he seizes one of these and waves it in the face of oncoming traffic, which is theoretically obliged to stop. Then, safe on the other side of the street, he deposits the yellow flag in another stand. A pedestrian thus carries his own red light, so to speak.

Stand near the Ginza viaduct and brace yourself. Overhead the "bullet express" of the New Tokaido Line shoots past. This, the fastest train in the world, reaches a speed of more than two miles a minute at several points, covering the 320 miles to Osaka in 190 minutes. Sixty trains of twelve cars each run every day in each direction and make the area a shrieking metallic shambles.[2] One subway station nearby, Tokyo Central, handles more than a million passengers a day; so does the main station at Shinjuku across the city. The local trains are steel cylinders packed with passengers like a duffel bag stuffed with live flesh; they operate in the rush hours far above normal capacity, and stout guards known as "pushers" stand on the platforms in order to squeeze people forcibly into the teeming cars. There are some 2,700 such pushers in Tokyo. The crush is so monstrous that passengers sometimes have their shoes torn off, or lose a sleeve off their coats. But in general the service is safe, swift, and almost frighteningly efficient.

Maybe the Hotel Okura has no smut, but Tokyo in the large is one of the most smog-laden cities in the world; seventeen tons of

[2] Jane Brody in the *New York Times*, November 14, 1966.

dirt, dust, and smoke particles per square mile per month fall down on it, and "Tokyo asthma" is recognized locally as a specific illness caused by contaminated air. I noticed in one newspaper an astonishing story about what was called a "bad air" intersection near a new loop road, where a sudden dip slows up traffic and makes smog accumulate. Here oxygen tanks have had to be set up and made available to elderly pedestrians and children, and the cops are restricted to two hours of duty at a time, although they inhale oxygen before taking up their posts.

More Items in Physiognomy

Cities have characteristic colors, I have always thought. Istanbul is blue and silver, London red, Marrakech pink, and Paris gray. Tokyo is the color of whitish cardboard or raw cement. A good many cities have distinctive smells as well—London of fog and wet leaves, Vienna of ozone, beer, and goulash. Tokyo has a faint, pervasive odor of fish fried in grease.

Most of Tokyo is built of wood, which, the Japanese say, is incomparably the best of building materials; wooden buildings exist here that are still serviceable after several centuries. Wood swells and keeps the heat in during the winter and contracts and lets in the breezes in the summer; above all it makes for quiet, and is beautiful.

Tokyo is one of the few great cities in the world (Rome is another) without a single skyscraper—although it has an imitation Eiffel Tower a bit higher than the one in Paris—and permission was recently given to build a thirty-six-story office building, so engineered that it will presumably be earthquake-proof. In general, both residential and commercial buildings have been restricted for many years to a height of 31 meters (101.7 feet), which means eight to ten floors; the average is only three. The principal reason why buildings have been so low is fear of fire and earthquakes.

Tokyo has fewer conspicuous parks than any great metropolis I have ever seen, fewer tennis courts, and fewer playgrounds attached to the schools. No room. (One reason why stairways in most Tokyo homes are very steep is to save space.)

One element in mystification which most foreigners find stupefying is that streets (with few exceptions) are not named, and house numbers, if they exist at all, are seldom displayed. Moreover, numbers do not go in sequence, so that No. 7 on a given street may be next to No. 26—it seems mad. The numeration depends on when the house was built. This is particularly puzzling to the occidental visitor because he has always been led to believe that the Japanese are above all a rational people, well organized, efficient. Yet the incredible fact remains that thousands upon thousands of streets, ranging from sumptuous boulevards to squalid crooked lanes, have no names at all, and people live in houses without any discernible identification. I have often wondered how the postman functions.

How is this possible? The best answer is that it just started that way when today's great cities were villages and everybody knew everybody else; the Japanese are extremely tenacious about modes and institutions. They have been wearing geta (wooden clogs) and digitated socks for a thousand years.

Much of the sheer demented fun of visiting Tokyo comes from the lack of identification of streets and houses. The visitor feels like a pinball eccentrically let loose on a board designed by a blind lunatic. A trip by taxi can be a spirited adventure. Invitations to Westerners are almost always, even from Westerners resident in the city, accompanied by a map with a text in Japanese; you show this to the hotel doorman, who, studying it carefully, shows it in turn to the taxi driver; then you set out, usually on a course marked by fierce, rapid zigzags. Some resident foreigners put up small wooden flags, like the owner's sign on a Westchester lawn, on the open streets to mark the route to their houses from a given point on the map. If, even so, the driver gets lost, he searches out the koban, or local police post, and asks for further directions. An animated conference, usually with several cops participating, follows, and is often lengthy. Tokyoites may have just as much difficulty as foreigners in finding their way about. I have it on good authority that about a hundred thousand citizens *a day* ask directions from the police!

When, finally, you do reach your destination, the driver triumphantly flips open his rear door by means of a gadget in front. I have never seen this alarming device in any other city. Mission accomplished. The driver smiles, and he doesn't expect a tip. Tokyo is just about the last tipless—well, virtually tipless—city in the world, money-conscious as it is.

Another baffling contradiction has to do with cleanliness. The Japanese are reputed to have a passionate addiction for being clean, and their baths, with steaming-hot clear water (you are not permitted to soap yourself in the tub), are celebrated. But only about 30 percent of Tokyo houses have baths (citizens go to bathing establishments), and at least 60 percent have no flush toilets. What is euphemistically known as "night soil" is collected in buckets, and these are put out at designated street corners where the garbage service collects them twice a week.[3]

Edoites are justly proud of their technical inventiveness and ingenuity. Their TV is first-class, and the cameras and watches they make are, as everybody knows, examples of the most robust, skillful artisanship. But nobody has ever succeeded in building a typewriter capable of handling the Japanese language satisfactorily. Typewriters with Japanese characters do indeed exist, but they have to be immensely cumbersome. Of course it is the formidable complexity of the language, built on a double set of phonetic symbols, on top of Chinese ideographs, which causes this difficulty. Try to look up a name in the Toyko telephone book! There are some five thousand ideographs, which cannot be alphabetized, in common use, and to track down a name is a job for a specialist. Three sets of coordinates have to be used.

Tokyo is a desperately serious city, but no capital on earth

---

[3] Much of this waste is shipped out on dredges and tossed into the sea, but some is transformed into usable fertilizer by a purifying chemical process. Recently a night-soil collector was caught in the act of dumping human debris into a nearby manhole instead of taking it to the disposal plant. A local English-language newspaper published the news of this under the headline, "A Dirty Stinking Trick."

rivals it for organized frivolity. It contains no fewer than 97 *thousand* restaurants, boîtes, bars, night clubs, joints, and other places of public entertainment. The catch is that most of these are tight shut by midnight, so that citizens can be fresh for the business of the next day, i.e., making money.

Root and Substance

Tokyo, correctly Tokyo-to, means "Eastern Capital." Formerly called Edo ("Mouth of the Estuary"), it does not have a particularly long history; it doesn't go back into time like Rome, Baghdad, or Peking. A feudal baron named Otaa Dokan set up a castle here on a marshy site near the Sumida River in 1457, and Tokyo has spread out from this well-chosen spot, where the Imperial Palace stands today, in roughly concentric circles ever since. Similarly, Moscow grew outward from the central core of the Kremlin.

Tokyo did not become the capital until the seminal historical event known as the Meiji Restoration in 1868, when it replaced Kyoto ("Western Capital"), but it had been a substantial city for many years. In fact, by the end of the eighteenth century, it was already the biggest city in the world, with a population around 1.7 million. It grew so fast from a fishing village to a metropolis largely because the shoguns, or military rulers, made it their headquarters. The monarch, secluded in Kyoto, was ignored.

Soon a trading class grew up in Tokyo (then Edo) to serve the daimio, feudal lords. These merchants presently became financiers, and rose steeply in power and influence. The country was opened up to the world in the middle of the sixteenth century, and industrialization eventually got under way. Some Edoites are called Eddoko today, which means that their families go back to the beginning of the modern period, and that they are third-generation residents of the city—a proud distinction.

A phoenix, Tokyo has several times risen literally from the ashes. There were devastating fires before the twentieth century. In 1923 came the great earthquake and fire which killed at least a hundred thousand people. This catastrophe came at exactly the worst moment of the day—at noon, when countless housewives were lighting fires in hibachi (small portable stoves), for the midday meal. A Tokyo house, being made largely of wood and paper, will burn like a torch if a window is broached; the earthquake broke the windows, and the conflagration followed. Then in 1945, toward the end of what Japanese historians call "the Pacific War," fleets of United States bombers burned Tokyo out. More than a third of all buildings in the city were destroyed. Great fire storms sucked oxygen out of the air, suffocating men, women, and children by the thousand; more people were killed in Tokyo—little known as this fact may be—than in Hiroshima and Nagasaki put together.

The local adage is that Tokyo citizens fear four things above all—fire, earthquakes, typhoons, and their fathers. While I was in the city recently, the embassy of a Middle Eastern country was burned to the ground because it was situated on a street so narrow that the fire trucks could not squeeze through. There are thousands of such buildings in Tokyo. The city has two or three minor earthquake tremors a day, mostly too slight to be felt; a jolt that shakes the wall once a month or so; and a real jolt several times a year.

An earthquake can be a terrifying experience because there is no telling how long the shock will last or how severe it is going to be. Many years ago I experienced one in Tokyo; what I remember most was the peculiar lateral oscillation of the floor, while tall posts outside bent almost double. The safest place in a house during a quake is under a doorway. Or lie flat under a heavy desk or table. Do not go out into the street.

✿

On a map Tokyo looks rather like a griffin. It is a knot of communities, not merely one, containing 14 "cities" or boroughs, 23 "towns" or wards, 26 villages, and 7 islands. In New York and London each borough or big division of the city has, speaking broadly, its own special distinctions, but Tokyo is more diffused;

each district seemingly contains elements of almost everything. The major department stores are not, as in most cities, focused in a single area, but are to be found in several neighborhoods, and at least three different principal centers for entertainment and the theater exist; there are three different Broadways. Similarly, there is no single fashionable neighborhood, although some, like the embassy quarter, are better than others. I asked time and again what was the Tokyo equivalent of Mayfair in London or the East Sixties in New York. There is no equivalent. The elite live all over the place; the richest of industrialists may, as I have already mentioned, have a hidden palace in a slum. I asked where the newly arrived manager of a large American corporation would choose to live, and the answer was, "Almost anywhere."

Even the casual visitor, once he gets over the initial confusion and adjusts himself to the fact that streets do not have names, except a few like the Ginza, will come to recognize three or four Tokyo districts with outstanding characteristics, although these may overlap. The GINZA, near the Imperial Palace, is both a street and a neighborhood; the former, originally known as the Street of the Silversmiths, goes back to the eighteenth century. It is a kind of Bond Street or Fifth Avenue, with sumptuous shops, but it also contains many elements of the humble—for instance, the equivalents of our penny arcades (pinball games are, incidentally, set up vertically in Tokyo to save space). Once, directly in front of an extravagantly fashionable store, which had opulent bowls of tropical fish as window decorations, I saw a ring of passers-by eagerly surround a storyteller on the pavement—like something out of Darkest Africa.

Nearby is AKASAKA (pronounced Aksaka), where the most powerful of the daimio lived, and which still contains remnants of the old feudal estates. Here are several of the leading hotels, embassies, and foreign restaurants; this is the haunt of geishas, connoisseurs, and up-and-coming businessmen. The old Yoshiwara, or prostitutes' quarter, was close by. A contrast is ASAKUSA, a torrentially crowded and colorful network of bazaar-like little shops on gnarled streets; sometimes this is called the poor man's Ginza or "the real Tokyo." It has an explosive night life, too. Still different is ARAKAWA in the north, which resembles the Jersey flats.

SHIBUYA, west of the Ginza, a prosperous middle-class area, has the Meiji Shrine and the magnificent new stadium and swimming pool built for the Olympics by Japan's foremost architect, Kenzo Tange. North is SHINJUKU, a huge submetropolis largely created since the war, with big department stores, a fabulously extensive and naughty amusement area, and hordes of riffraff. But ROPPONGI comes closer to being Chelsea or Greenwich Village. Here beatniks roam.

The biggest wards are OTA to the south and SETAGAYA to the west, each with 750,000 people. BUNKYO-KU, not far from the palace, is the seat of Tokyo University, one of the great educational institutions in the world. One nearby district, KANDA, resembles streets near the British Museum in London, crowded with rows of bookshops. The local Bowery is called SANYA, and manners here can be very rough.

Despite this variety of neighborhoods, Tokyo is remarkably homogeneous. It has no Harlem or Watts. There are only two "minority groups," the Koreans, who live mostly in a kind of ghetto of their own, and the Burakumin. For reasons not clearly discernible, Koreans are disliked, feared, and looked down upon. The Burakumin comprise a special social category something like the untouchables in India; the prejudice against them is largely occupational, since they are mostly garbage collectors, slaughterhouse workers, and the like. Feeling about them is so hostile that a word, "Eta," often used to describe them, is not said aloud in polite society. But some Etas, no matter how they are discriminated against, manage to rise these days. The former rector of an important university is an Eta, and so is the president of a large manufacturing company.

About 10 percent of the population of Tokyo speaks English more or less, and the city has five daily English-language newspapers. "Batman" and "L'il Abner" appear in the local comic strips, and I saw American movies and golf lessons on TV. American-style baseball (*Beisu-buru*) is by far the most popular sport, and bowling is a rage. The baseball teams are organized into leagues, as in the United States, and a world's series on the American model transfixes the city every autumn. Half a dozen American big-league players have moved to Tokyo and play professionally for local teams.

The city has, on a broader level, resisted Americanization more than accepted it, but signs (very small signs) exist in English almost everywhere, like "Subway," "Exit," and "Road Under Construction" (they all seem to be). Many establishments carry English names, some of which are not quite idiomatic, like "Ladies' Fitness Center." I saw "Ace Pig Lard" on a butcher's cart, and shops called "Charm Salon," "Brother Service Center," "Fruit and Flower," "Fancy Corner" (confectionery), "Fun for Living" (furniture), "Mr. Ship," and "Nutty Chocolates." Coffee houses are named "Dreama-Tone" and "Doncamatic." I bought a pencil with an electric light attached, the label of which said, "World-Recognized Masterpiece With Seven Patents—Just Putting Cap on Barrel End Lights It Pat." But one of the most extraordinary features of Tokyo is that the long American occupation after World War II left so little trace. Walk ten steps off the Ginza and you are deep in an Asian jungle.

Some Sights and Things to Do

One suggestive phenomenon in this remarkable amalgam of a city is that its principal "sight," the Imperial Palace, cannot be seen—anyway, not in the ordinary course of events. Some people do see it, of course, but not the casual tourist, because, although it commands the center of Tokyo, it lies behind heavy glossy walls, nine feet high, and a wide, briskly flowing moat. Nearby a sight that can be seen, and which is visited by thousands, is the Yasukuni Shrine dedicated to Japanese soldiers killed in battle, every one of whom becomes a kind of god in the country's intricate and crowded pantheon. Worshipers approach, clap loudly to summon up the spirit of the dead warrior, and—presumably to pay for this privilege—deposit a small coin in a receptacle. Another sight, as crisply modern as the shrine is archaic, is the new Roman Catholic cathedral built by Kenzo Tange, than which no church—with its sharp angles, steep tower, and interior cement walls—could be more spectacularly unorthodox. It contains no direct pictorial representation of Christ at all, and the Virgin is restricted to a position in the courtyard.

The Ginza quarter is marvelous at night, with electric signs and advertisements vibrating in gaudy colors. I was reminded of a famous remark by G. K. Chesterton about Times Square in New York: how doubly stunning the sight would be for a man who could see but not read. Few Westerners can understand the Japanese ideographs which charge back and forth in brilliant neon, but this makes them even more fantastic, exotic, and tantalizing. Otherwise Tokyo hasn't much to offer in the way of standard "sights." There is no Tower of London, Vatican, Louvre, Uffizi, or Bridge of Sighs.

The city is, nevertheless, crammed with its own kind of marvels. One day we saw a judo-karate exhibition. A sixty-nine-year-old "red belt man," or senior champion, tossed boys around as if they were bags of peanuts—boys forty pounds heavier and half a century younger. One was one of his own sons. Two women, one armed with a sword, one with a spear, put on an extraordinary duel, almost a dance. Then there were dramatic demonstrations of the methods whereby a man may protect himself, with bare hands, against an assailant with gun or dagger. The border line between judo and karate is something for purists to argue about. The best definition of judo I ever heard was that it was "body chess"—a game of wits, anticipation, and throwing the opponent off balance. As to karate, we watched with incredulity boys punch through five layers of wood, crunching them as if they were paper; they did this with their feet too, even without a firm stance on the ground.[4] The trick is speed, not power. And

4 Of course karate exhibitions have lately become commonplace all over the world.

we learned that, incredibly enough, a trained karate expert must register his actual hand with the police as the equivalent of a deadly weapon, which is exactly what it is.

A visit to a great Tokyo department store, like Takashimaya or Mitsukoshi, the "Macy's of the Orient," is revealing. The top floor is used as a rule for cultural exhibitions of handicrafts and the like, and doll-like pretty girls in kimonos (they look like mannequins), bow to the visitor at each escalator stop. These huge stores are, like Tokyo itself, adding machines. Prices ranged from the unbelievably low to the substantially high. I saw several kimonos priced at 100,000 yen ($277.78), and I was told that a really good antique obi, the kind of sash which looks like a parachute pack attached to the back, can cost 500,000 yen, more than $1,300.

Then, too, the beer halls are worth attention. I went to one called the Sapporo on the Ginza, which calls itself a "health center" as well. A sign outside says in English, "SANITARY CONDITIONS IN THIS SHOP," and a price list advertises a sirloin steak for 800 yen ($2.23), an American "Clubhouse Sandwich" for 300 yen (84 cents). Though the menu was in English as well as Japanese, I did not see a single Westerner in this whole large, ambitious, crowded establishment. The atmosphere was extremely decorous. Men sat alone with serious expressions, intently silent and drinking beer slowly out of steins as big as those in Munich. Price of a stein—250 yen or 70 cents.

Tokyo has an immense number of coffee shops, most of which are tiny, because land is so expensive and rents so high. But one which I visited, the Chopin, is gaily decorated, with purple lights, fills four stories, and serves music as well as coffee.[5] This is a familiar Tokyo phenomenon. Many coffee shops have large libraries of Western recorded music, to which music-lovers go as if they were going to a concert. You order a cup of coffee (100 yen), and the waitress brings a large album, or catalogue, listing

[5] Once more I must mention how frustratingly difficult it is to find anything in Tokyo. It took me three tries to locate the Chopin, although it is well known and I was accompanied by a Japanese interpreter.

the records available—in the case of the Chopin, more than four thousand. I saw that Mozart had nine pages, Beethoven eight. Then you fill out a slip of paper listing your requests, and await your turn to hear them. No drinks are served, and the atmosphere is serious; clients whisper an indignant "Shh" if you talk in the middle of a recording. Cost of the music: nothing.

The Tokyo theater offers anything from a fairly good imitation of the Rockettes to classic Noh, with modern plays, strip-tease, and puppet shows thrown in. This is a city both knowledge-able and sophisticated about things theatrical; the revolving stage has been used here since the eighteenth century. The new Imperial Theater, where we saw a stunning Kabuki performance, has just been completed, and is comparable to Lincoln Center in New York. A vibrating metal plate is set into the pavement just outside; this is to shake dirt off your shoes, which do not have to be removed. Inside there are several gadgety innovations; we visited the star's dressing room, which is five or six stories above the stage; actors are cued by television. Incidentally, we never saw more than two or three foreigners at any Noh or Kabuki performance, although the theaters are large and thronged.

I found Noh heavy going, but my wife survived—and liked—three different performances. A single act can last three or four hours. Sometimes performances start at eleven in the morning and go on all day; a Japanese family will make a picnic of the occasion. Kabuki, which combines pageant, drama, elaborate stage effects, colorful costumes, formalized dancing, and some of the most brilliant acting anybody ever saw, is, if you are properly attuned, romantic beyond measure—even if some of the music sounds like an iron pipe being clobbered with a stick. Two of the most glorious and sensitive actors in the world—incontestably—are the Kabuki players Shoroku and Utaemon.

Movies are big business in Tokyo, and several Japanese films have become well known everywhere for their artistry and emotional punch. We were lucky enough to have several meetings with Yukio Mishima, who is not only one of the most distinguished Japanese writers of his generation but an accomplished actor, athlete, theatrical director, and movie-maker as well. One of his recent short films, *Patriotism,* is based on a case of ceremonial hara-kiri, and is a masterpiece of explicitly stated love, horror, and soaring tragedy. It has never been shown in America.

What Else Does Tokyo Offer?

About restaurants I could write pages. What makes them so fascinating is the ceremonial attached to dining out. You reach your destination, which may be in a dark, shabby cul-de-sac, and suddenly find yourself, behind the façade, in a nest of gleaming, fine-grained, wood-and-paper boxes. The patron bows; the servants bow; everybody bows, to the accompaniment of chirps and titters. You take off your shoes, and put on slippers provided by the house; if you go to the powder room later, you are given a different pair of slippers. You are led to a chosen room, and sit down on flat cushions at a low oblong table or a series of tables put together; there is no separation of tables no matter how big the party. The problem is what to do with your Western legs. In a few restaurants in Europeanized hotels the Tokyoites have solved this problem by the ingenious device of having a concealed well, or pit, under the table. So you can drop your legs down even though sitting on the floor.

Parties generally number a minimum of six. It would be unthinkable to dine alone. In a really good Japanese restaurant no group of diners ever sees another group, like patients in a psychiatrist's office in New York. Each party has its own exquisite and tranquil room, and each is private. There is no printed menu, and the host will have usually ordered dinner in advance, but, after the meal is over, the guest may be given a fan or other souvenir recording what dishes he has had. The dining hour is early, around seven, and dinner in a classic restaurant can be definitely expensive, $50 a person or even more.

A hot, damp washcloth is given each guest to begin with, and the pretty waitresses with their stiff coiled hair kneel behind you, prepared to assist the Western barbarians who do not understand truly civilized and cultivated oriental ways. Chopsticks are made of light wood (not ivory, or plastic, as in China) and are not completely separated, in order to prove that they have never been used before. After each meal they are simply thrown away. If you must have a fork, it can be requested. I never saw a knife.

Monday is the big night, as it is in bars and night clubs, because the average Tokyo husband, having undergone the stifling rigors of a weekend at home with his family, is doubly impatient to go out on the town. Japanese *wives* are practically never to be seen in elite restaurants. A man seldom takes his wife out in company. At an elaborate dinner the waitresses belonging to the restaurant may be supplemented by geishas, but these cost a great deal; the normal geisha fee is 10,000 yen ($28) per hour, and these charming young ladies calculate their hour at fifty minutes. They help serve food, play little games with the guests, pour sake, dance, and in a highly stylized way fulfill their function, which is to entertain and be agreeable. Once at a large garden party our host had at least twenty geishas, and they looked like silver butterflies floating over the lawns and near the flowing brook, shimmering, evanescent, light as pale flames, and just as pretty as they were exotic.

Some Japanese food may seem strange to our palates, but most of it is delicious—not only tempting to eat but to look at, because of the artful way dishes are composed. Almost everybody knows about tempura (seafood and other substances cooked in a light batter), tofu or bean curd, raw fish, and sukiyaki, correctly pronounced "suk-yaki." But not everybody has heard about Japanese beef. The best quality comes from cattle which have never been permitted to have any exercise and which, indeed, never leave their stables; they are fed special grains soaked in beer and are actually massaged by their owners so that their meat is of the tenderest—so marbled that it looks like dictionary paper.

A full-scale formal Japanese dinner is copious as well as elaborate; we went to one at which there were thirty to forty different *plats* in nine major courses. First, in a narrow trough of bamboo, came the hors d'oeuvres—trouts' eggs, which looked like fig seeds, squid, a tiny arc of lime, and crushed chrysanthemum leaf. Next appeared a small hibachi, or charcoal stove, one for each guest, and a pot in which your waitress helps you boil mushrooms, a rice patty, and a starchy dish resembling gnocchi, but more delicate. Shredded bonita came next, accompanied by red beans, sesame seeds, and a sauce hot enough to tear out your tongue; after this were salmon roe, slim stalks of zucchini, raw carrots cut into the shape of maple leaves, tiny liver balls light as gauze, crab, ginger, chestnuts. Meanwhile came the *pièce de résistance* —raw tuna, bream, and other fish. Paper-thin slices of wild duck followed, served with a blackish kind of seaweed. Then came the tempura course with six or seven varieties of seafood accompanied by a hot sauce made of chives, together with chrysanthemum fried in oil. We progressed at last to slim delicate slivers of filet steak, grilled before us, together with turtle soup, rice, papaya, and a very sweet custard made of beans.

It was not surprising to hear that the Imperial Palace often sends over to this restaurant for snacks.

To my mind, one remarkable thing was that European as well as Japanese food is astonishingly good in Tokyo. The oysters are the best I have ever had anywhere, and the green grapes, which seem to be actually perfumed, are almost as big as plums— incomparable! I had the best Wiener Schnitzel I have ever had outside Vienna in the international restaurant of the Okura, and the best mixed grill as well. Chinese food is also excellent in several places in Tokyo.

A dinner like the one I have just attempted to describe is, of course, an occasion, but we went to several others almost as festive and elaborate with Japanese friends as our hosts. A second category of restaurant is the counter bar, much less expensive, which specializes in tempura—maybe twenty different kinds. You sit on a stool at a long wooden bar, as a counterman dips various bits of fish and meat into the boiling batter. Then, too, there are steak houses where your beef is cut into cubes which you yourself grill to your taste on a large square hot plate, known as a teppan, set into the center of the table. And there are thousands of lesser restaurants. You don't have to be a millionaire to enjoy restaurant life in Tokyo. Twice I had simple one-course dinners —fried rice—which only cost a couple of hundred yen.

The standard Japanese drinks are sake, which is usually served warm, beer, and a local Scotch called Suntory. Sake, which is put at your disposal continuously during a ceremonial dinner, is brewed from rice, contains between 12 and 14 percent alcohol, and can be unexpectedly potent. The city has several fascinating stand-up bars which offer the client a variety of sakes out of old casks, like sherry in Spain; at one, while we sipped, we nibbled periwinkles. Mixed drinks of bizarre composition are a local speciality. You see cardboard or plastic models of these outside the raucous little bars in Shinjuku and elsewhere. In actual liquid form I encountered the Blue Moon cocktail which, implausibly enough, consists of dry gin, *crème de violette,* lemon juice, and fresh mint, and the Paradise cocktail, containing apricot brandy, gin, orange juice, and something known as "gum syrup."

As to night life, the number and variety of bars and boîtes is, as I have already mentioned, staggering. There are short streets containing forty bars in a row, side by side; never in my life have I seen so many bars to serve every taste. They range from grandiose multifloored establishments with elaborate floor shows and a thousand (literally) hostesses to dank little holes in the wall, as scummy as chewed gum. Outright prostitution is technically illegal, with the result that myriads of girls disguise themselves as "hostesses," and semi-pros swarm everywhere; they have come close to putting the geishas out of business except for the very rich. Prices at the good bars can be exorbitantly high. My wife and I took two guests to a fairly respectable "club"; the bill was the equivalent of $48 for 48 minutes, though I do not think that the total amount was calculated by the minute. Two or three pretty little opaque-faced hostesses attended us, but we did not have more than seven or eight drinks in all, and nothing at all to eat. But I was assured that I had got off lightly. There are places at which the cover charge alone is $27, and plenty of Japanese businessmen think nothing of spending a thousand dollars on an evening.

Where do they get the money? The answer is that they don't. Their offices pay. The extravagances of night life are made possible by the expense account, the use of which reaches its apogee in Tokyo. Both employers and employees, in order to avoid taxes, find it more convenient all around for salaries to be low and expenses high. Bills in restaurants and night clubs are never itemized and, in fact, are seldom even signed by the customer. He simply gives the name of his company, and the establishment mails the bill to the employer the next day. And, of course, the bars and restaurants are, in theory at least, tipless. Cash seldom passes, except in the form of *douceurs* to the hostesses, and these are not obligatory. However, there do exist some few Tokyo residents who do not have expense accounts, but even these manage to live up to the universal maxim, "Earn money by day, spend it at night." To aid them an organization has been set up known as "Tokyo Lease," according to a recent dispatch to the *New York Times*. If you are out on the town and run out of cash, this useful agency will dispatch a "Cash Ambulance" to rescue you at any hour of day or night; there are about forty calls a night. Money is doled out to the necessary amount, and interest charged at 3.8 percent a month. Security is nothing but your calling card.

Fundamentals

Tokyo politics are a mishmash, a stew. The city, which is also a prefecture (province), is governed by a Metropolitan Assembly of 120 members and an elected governor (mayor), who serves a four-year term. Interrelations with the national government for roughly 40 percent of its budget; similarly—but not in the same proportion—New York City is financially dependent on New York State.

Dr. Ryutaro Azuma, the Governor of Tokyo when we were there, is almost seventy, but, like so many Japanese, he looks much younger; brisk, alert, smooth-skinned, he could pass for fifty. A physiologist by profession, he held an important post in the Ministry of Health for many years, and he has been a professor at the University of Tokyo since 1946.

Tokyo, Dr. Azuma told us during a lively interview, is 500,000 units short in housing. Traffic is, as is only too obvious, another compelling problem. A ten-year plan was set up in 1961 under the supposition that the city would have a million automobiles within ten years, but this figure was reached in five. There has been a considerable growth in crime and juvenile delinquency. But what has troubled Dr. Azuma most during his two terms as Governor (he was first elected in 1959) has been political strife based on a series of noisy scandals. Several members of the Metropolitan Assembly were accused of corruption last year, the speaker was arrested, and seventeen assemblymen were indicted on various charges involving graft. Dr. Azuma himself was not directly implicated, but the cry was strident for his recall.

The scandals arose partly out of the peculiar but characteristic Tokyo habit of gift-giving. Twice a year, on stated dates, it is the accepted custom for almost everybody in a professional relationship with another to give him a present. Then, too, when a person dies, the surviving widow or widower receives a gift from friends of the family or relatives, usually money, as a traditional token toward paying for the funeral, no matter how rich he or she is. Even the Emperor gives such gifts, and members of the imperial house receive them.

Suppose you are a contractor doing business with a government official, or even a performer working with a TV agent or producer. Gifts are expected and duly sent. Suppose, then, that the government official involved, or the TV producer, hints that he has no need for a case of imported Scotch, a membership in the country club, a new camera, or even an automobile, but—whisper it—would prefer the equivalent in hard cash instead. So the gift goes out in the homely shape of yen.

In the case of the Tokyo Metropolitan Assembly, the speaker and other members were accused of distributing several million yen to buy political support.[6] The speakership is an eagerly sought post because a large expense account goes with it. The upshot was that the Assembly had to be dissolved in the summer of 1965, to the tune of screams of scandal, and new elections held. These resulted in an important shift in the political line-up in Tokyo, a matter gravely embarrassing to Dr. Azuma and the national government.

Dr. Azuma belongs to the Liberal-Democratic (conservative) Party, the same party which, under Prime Minister Eisaku Sato, runs the country as a whole. The Liberal-Democratic representation in the Metropolitan Assembly fell from 66 to 38 in these 1965 elections, as a result of indignation over the scandals, which also touched the national government, and the Socialists, the principal opposition party, rose from 32 to 45, thus becoming the biggest single party. Another socialist group which had had no seats at all won 3, and the Communists rose from 2 to 9. Left-wing parties, in other words, won a plurality, though not a majority, in the Assembly.

More than this, the Komeito (Clean Government) Party rose from 17 seats to 23. This is the secular arm of a remarkable Buddhist reformist movement known as Soka Gakkai, which has become a major political as well as spiritual force all over Japan, although it is difficult to know precisely what it stands for.[7] After a preliminary period of supporting Azuma, the Komeito grew to hold the balance of power in the metropolis and could tip it either way toward the conservatives or Socialists. But when I was in Tokyo, it had become reluctant to take sides, nor would it formally enter the turmoil and clangor of domestic politics by running a candidate itself, although its membership is huge—said to be one out of every six adult citizens in the city.

Then came new elections for the governorship of Tokyo in April, 1967. The winner was Dr. Ryokichi Minobe, who became chief executive of the city by a narrow margin over the conserva-

---

[6] *New York Times*, April 18, 1965.

[7] Incredibly enough, there are said to be 70,000 different Buddhist sects in Japan.

tives. The Komeito contested this election, but ran third, although Soka Gakkai itself is potentially the most powerful pressure group in the nation. Dr. Minobe, who is sixty-four and who had been professor of economics at Tokyo University for some years, is an emphatic man of the left, and won by reason of the support of various left-wing groups, including the great trade unions and both the Socialist and Communist parties—also by a large share of the women's vote. Minobe has been an athlete, farmer, and newspaperman as well as a professor and government official, richly erudite in several fields and clement of disposition. Back in 1938 he spent eighteen months in jail as a suspected Communist, and he is probably more to the left than any mayor of any great city in the world, the Communist capitals excepted.

Tokyo: What Runs It

1. Business and commercial interests like the Mitsuis, as well as members of the newer school of industrialists such as the men who vigorously run Sony (radios, et cetera), Honda (motorbikes, automobiles), and several immensely wealthy shipowners. As to Mitsui, which is older than the Bank of England, it is still the world's largest trading company and handles 11-12 percent of *all* Japanese foreign trade.

2. The old men, like the Genro of prewar days. Longevity is still venerated, and respect for old age (*Keiro-Shugi*) is universal. The power of several wily old advisers to the government, whom nobody outside an extremely limited circle may have ever even heard of, is both mysteriously cloaked and practically substantial.

3. The women.

4. The bureaucracy.

5. Religious or semireligious influences, as represented by Soka Gakkai and Komeito, which may reflect a growing reaction against materialism.

Beyond this it is difficult to speculate because such an abstraction as "the voters" means little in Japan, even though it operates under ostensibly democratic processes. Power is still invisible.

But there ought to be a line at least about two weighty organizations in the realm of big business—the Kei Dan Ren (Federation of Economic Organizations), which corresponds roughly to the National Association of Manufacturers in the United States, and the Nikkeiren, or Japan Federation of Employers' Associations, which enrolls some of the best brains in the nation.

I asked almost everybody another question: what does Tokyo need most? The most laconic and suggestive answer I heard was: "Fewer people." Another was: "A realistic price system."

A Miscellany of Singular Qualities

Tokyoites will, it seems, go to almost any lengths to avoid sunburn. Let so much as a single shaft of pale sun penetrate the trees around the Meiji Shrine, and startled strollers will put handkerchiefs over their heads and faces or dive for cover. Practically all citizens make constant use of calling cards, and the visitor would be well advised to buy some himself, printed in English on one side, Japanese on the other. Almost everybody is a camera fiend; Adlai Stevenson once made the quip that the whole country was a "photocracy." Another piquancy is that no Japanese has a middle name.

Defeat in World War II was a traumatic shock of the first degree to most Tokyoites (Japan had never lost a war before), and a sociologist might find some significance in the fact that the three most popular brands of cigarettes today are named Hope, Peace, and Pearl. During most of the MacArthur occupation the Japanese seemed dazed. Now, feeling that they were betrayed by their own leaders, and to save face, they want to scrub out almost the whole of their militarist past. Moreover, they know well that peace pays. When I asked a Tokyo friend if he resented us, the Americans, for having beaten his country, he looked at me with astonishment, saying, "Of course not—we admire you!"

Elements in the Japanese character are hysteria, punctiliousness, solidarity, stoicism, industriousness, tenacity, and a tendency to conform. The Japanese do not complain, but accept. Nothing is ever spontaneous in Tokyo, and nothing is ever lost. The intense politeness of the Edoite is, some psychologists say, a subconscious device to cover aggressive instincts. Nobody ever likes to take exclusive responsibility for anything, which means among other things that Tokyo is a city dominated by go-betweens. No single officer of a bank is ever responsible for a loan; no policeman is likely to give you a specific direct answer to a question. He will say, "One moment, please," and consult his comrade. Favorite phrases are *"Komari masu"* ("I am troubled") or *"Shikata-ganai"* ("It can't be helped").

The changes I noticed most in Tokyo in comparison to postwar days were mainly in realms of the physical. Young people are taller, better built, longer in the thigh, with better teeth, partly as a result of the new habits of diet which came in with the American occupation. In the old days mothers suckled their children for year after year, but this practice, which probably played a role in causing so many Japanese to have pursed lips, has pretty well gone out. Children no longer grow up bowlegged as the result of being made to squat for hour after hour on the floor. Few adults hiss when they talk, in the old sibilant manner. Many do, however, still wear flu masks on the streets. Eye care is better. Thousands of Japanese young women go these days to plastic surgeons who, by a simple operation, make their eyes straighter.

But what does all this add up to? What does the mad scramble of contemporary Tokyo signify? One notable fact is that there has been such a stupendous development with so little dislocation; another is that this is, by and large, a happy city. Welding of its several components proceeds methodically like some vast exercise in counterpoint. Tokyo is on the make, on the go—confident, practical, fantastically busy, and successful. Not long ago a great airliner smashed into Mount Fuji, and everybody on board was killed. Most passengers carried watches, and it was discovered that every Swiss, American, British, French, or German watch had stopped dead at the moment of impact. But every Japanese watch survived the crash and was running fine. Time does not stop for this great city in a hurry.

IN CONCLUSION

I don't think that this book needs a formal summary or count-up at any length. The treatment of each city carries its own particular conclusion, if only because each differs so strikingly from the others.

Nevertheless these twelve cities lie in a common frame, and point up the fact that the stupendous, terrifying growth of the contemporary urban metropolis confronts us with problems not merely vexing, as suggested in my foreword so many pages past, but perhaps insoluble. The European and Asian capitals are not paradises by any means, and are not likely to become so. They are not exempt from bristly troubles. Even so, we may perhaps learn something from their example, and thus be able to deal more effectively, more hopefully with our own failures and predicaments.

Moscow, with its spotless streets, is far cleaner than New York; it is a great industrial city, but has no smog. Warsaw is a more harmonious community from several points of view than Los Angeles or Dallas. Tel Aviv, Berne, and Bruges are more adequately administered than most American cities in an analogous category. The gross physical squalor, dilapidation, noise, and filth in the outskirts of several important American cities would not be tolerated in Stockholm, Ankara, or even Madrid. No city in Europe has an entrance so brutally unkempt as, let us say, Bruckner Boulevard in New York. Even Beirut could show Detroit or Philadelphia a thing or two, and Jerusalem, even if it is an oriental city, is better run and less offensive to the eye and ear than Pittsburgh or Atlanta.

Another factor to mention is that good civic health obviously does not lie merely in such realms as the birth rate or the merciless impact of whirling rubber on asphalt and concrete. Public spirit, urban morals, count as well—perhaps more. Here, too, the Europeans are far ahead of us. Of course corruption on the municipal level exists in Europe and Asia—consider Rome or Tokyo—but on nothing like the monstrous plumed scale of several American cities. Citizens are law-abiding to a degree, and they have pride in their civic environment. Brussels, as an example, if I may repeat a word from my text, had exactly three murders last year, of which only one was premeditated; Chicago had 552. And I might reiterate that in several European capitals—London and Paris, for instance—the elected officials who serve the municipality work without pay. Politics do not go by patronage. The police are honest by and large, and a crooked judge is virtually unknown. Little exists of the familiar American tendency to outrageous, promiscuous violence in schoolrooms or city streets. There have been no conspicuous political assassinations for many years. Then, too, Europe has never had political bosses, satraps, tetrarchs, of the type all too grotesquely conspicuous and vulgar on the municipal level in the recent American scene—no Pendergasts (Kansas City), no Curleys (Boston), no Hagues ("I am the law!"—in Jersey City). We, too, must learn to care for our cities, if only so that they will take better care of us.

INDEX